SECURITY OPERATIONS

AN INTRODUCTION TO PLANNING AND CONDUCTING PRIVATE SECURITY DETAILS FOR HIGH RISK AREAS

SECURITY OPERATIONS

AN INTRODUCTION TO PLANNING AND CONDUCTING PRIVATE SECURITY DETAILS FOR HIGH RISK AREAS

BY

ROBERT H. DEATHERAGE JR.

 TURTLE PRESS   SANTA FE

To contact the author or to order additional copies of this book:
Turtle Press
PO Box 34010
Santa Fe NM 87594-4010
1-800-77-TURTL
www.TurtlePress.com

ISBN 978-1-880336-98-4
LCCN 2007047484
Printed in the United States of America

10 9 8 7 6 5 4 3 2 1 0

Library of Congress Cataloguing in Publication Data

Deatherage, Robert H.
Security operations : an introduction to planning and conducting private security details for high risk areas / by Robert H. Deatherage, Jr.
     p. cm.
ISBN 978-1-880336-98-4
1. Private security services. 2. Security systems. 3. Corporations--Security measures. 4. Crime prevention.
I. Title.
HV8290.D43 2008
363.28'9--dc22
                    2007047484

CONTENTS

INTRODUCTION   11

CHAPTER 1:WHY A PSC PROVIDER IS HIRED   13
    Protection Strategies   15

CHAPTER 2: PROVIDING PROTECTION   15
    Roles of a Primary Objectives   17
    Fundamentals   17
    Areas of Protection   18
    Threat Levels   18
    Who is the Client?   21

CHAPTER 3: THE CLIENT   21
    Ethical Considerations   22

CHAPTER 4: COMPONENTS OF PROTECTION   23
    The Bodyguard   23
    The Bodyguard as Part of a Protective Detail 23
    The Stand Alone Bodyguard   25
    Advance Work   26
    The Personal Security Detail   29
    Equipment Concealment Problems   29
    Low Visibility Vs. High Visibility   30
    The Close Protection Team   30

CHAPTER 5: COMPOSITION OF THE SECURITY DETAIL   33
    Terms And Definitions   35

CHAPTER 6: KNOWING THE THREAT   37
    Range of Threats   38
    Internal Considerations   39
    External Considerations   39
    External considerations include: 39
    Needs Of The Threat   40
    Threat Operations Cycle: 41

CHAPTER 7: RISK MANAGEMENT   43
    Risk Assessment and Reduction   44
    Threat Assessments   46
    Vulnerability Assessment   47
    Risk Assessment   48
    Conducting a Risk Assessment   48

6

CHAPTER 8: THE ADVANCE   49
    Responsibilities Of The Advance   50
    A Sample Advance Questionnaire   51
    Pre-advance Work   51
    Itinerary   51
    Time Constraints   53
    Time Requirement Standards   53
    Location   54
    Threat Levels   55
    Site Specific Requirements   56
    Other Considerations   57
    Keys To Successful Advance Work   58

CHAPTER 9: SURVEILLANCE DETECTION PROGRAM   61
    Types Of Surveillance   62
    Methods Of Surveillance   63
    Static Surveillance   63
    Long Term Static Surveillance   64
    Foot Surveillance   64
    Mobile Surveillance   65
    Technical Surveillance   66
    Aerial Surveillance   66
    Anti-Surveillance Techniques   67
    Counter Surveillance   69
    Setting Up Counter Surveillance   71

CHAPTER 10: AWARENESS LEVELS FOR THE SECURITY PROFESSIONAL   73

CHAPTER 11: SITUATIONAL AWARENESS   75
    Tactical Mindset   78

CHAPTER 12: GENERAL GUIDELINES FOR PSD OPERATIONS   79

CHAPTER 13: FOOT MOVEMENT TACTICS AND TECHNIQUES   81
    One Man Detail   83
    Two Man Formation   84
    5 Man Formation (Modified Diamond)   85
    Foot Detail (Open)   88
    Foot Detail (Close)   89
    The Security Detail's Mission   90

CHAPTER 14: PUBLIC VENUES AND FUNCTIONS   91
    Public Speaking Venues   92
    Reception Lines   93
    Entering Elevators And Enclosed Spaces 96
    Approaching an Elevator   97

Entering an Elevator 98
Inside an Elevator 99

CHAPTER 15: ROUTE SELECTION 101
Route Planning Principles 103
Selection Of The Route 105

CHAPTER 16: THE PHASES OF ROUTE PLANNING 105
Organizing The Movement 107
Hostile Environments/War Zones 107
Identify The Routes 111

CHAPTER 17: MOVING THE CLIENT 111
Terms And Definitions 112

CHAPTER 18: MOTORCADE OPERATIONS 113
Vehicle Embus And Debus 115
Embus Considerations 115
Debus Considerations 116
Reacting To An Attack 117
Considerations 120
Things To Remember 120
What Could Go Wrong 121
Choosing The Right Vehicle 122
Armoring Vehicles 123
High Profile Vehicles 128
Convoy Equipment 131

CHAPTER 19: CAR OPERATIONS AND TACTICS 133
2 Car Operations And Tactics 134
Making A 2 Car 136
Traffic Circles 138
Vehicle Reception 139
2 Car Moving Roadblock Escape 140
2 Car Stationary Road Block 141
3 Car Operations And Tactics 142
Making A 3 Car Turn 144
Lane Changes 146
Passing Through Intersections 148
Vehicle Reception 150
Reacting To Attacks 152

CHAPTER 20: THE COUNTER ATTACK TEAM (CAT) 153
Mobile Operations Procedures 155
Static Site Operations 156
Principles Of Use 159

8

CHAPTER 21: COUNTER-SNIPER OPERATIONS   159

    Counter Sniper Equipment   160
    The Counter-sniper Team   161
    The Firing Position   161
    Control Of The Cs Team   162
    Reporting SOP   162
    Location Reporting   162
    Reporting Individuals   163
    Reporting Vehicles   163

CHAPTER 22: BUILDING SECURITY CONSIDERATIONS   165

    Outer Layer Security   168
    Inner Layer Security   168
    Deliveries   168
    Conduct   169
    Visitors   169
    Random Security Checks   169
    Other Security Details   170
    Night Shift   170
    Fire Safety   170
    Indirect Fire   170
    Bomb Threats   171
    Temporary Site Security   172

APPENDIX   173

    Types Of Surveys   173
    General Site Survey   174
    Example Facility Checklist   177
    Remain Overnight Hotel Survey   184
    Airport Survey   188
    Hospital Survey   190

    Basic Physical Security Checklist 190
    Options For Vehicle Security   193
    Options For Perimeter And Access 194

    General Search Techinque   195
    Quick Body Search Or Frisk   195
    Detailed Body Search   196

    Route Survey Format   197
    Protection Orders Example   199
    Security Detail Operations Orders 199
    Observation   205

ACKNOWLEDGEMENTS

Special thanks to Cynthia my editor for all her work, to the graphics department for doing such a great job and to Keith Livingston, Mark Carlson, Rob Vaughn, Mike Eckard and John Taylor.

INTRODUCTION

With many companies from around the world now working in hostile, non permissive work environments and war zones, it has become necessary for them to be able to protect their people, equipment and locations so they can perform the work they are trying to do. Most companies contract out their security since it lowers the cost to them and allows them to set specific needs that a security company will have to meet so they, the client, can function. There are all types of security that a security company can provide but the three most common are protecting facilities, protecting people, and providing a secure means of transportation. Most of these areas are direct offshoots of the Executive Protection or VIP security industry and many of the tactics and techniques will be similar.

This book will talk about private security details, personal security details and/or private security companies, whatever you want to call them. It will discuss what type of training, experience and Standard Operating Procedures (SOP) they should have and the type of tactics and techniques that are used for different types of protection operations. While these Tactics, Techniques and Procedures will not be the exact same for every company, the basic concepts of personal security operations or very important person (VIP) details will always be the same regardless of experience, training, nationality or background.

Remember that Personal Security Companies and the PSDs they hire out for high threat areas and war zones are a fairly new phenomenon around the world. Security has taken a much higher place in the daily operations in those areas in which many government agencies, civilian service and construction companies now find themselves working.

PSDs and static security forces are the major part of the services offered by many of the new Private Security Companies that have sprung up to take on security jobs in high threat areas that the US military and other government organizations and companies simply cannot do, either because of manpower shortages or lack of experience and training.

Personal Security Details (PSDs) will do various things for the client but they are always supposed to have their primary objectives in mind: protect the client and keep him safe from harm, and protect the client's equipment so that the client can do his job. This means that all PSD personnel will be in harm's way quite often, but that is what they wanted to do when they took that job and why they were hired. For a client with the need to work in high threat areas, security allows them to do their job. If for some reason the client that hired a security company cannot do his job then he will not be able to finish his contract, which means that you as a security provider did not do your job.

A good security provider needs to remember that the people they are hired to protect have jobs to do also; they must have the necessary materials to accomplish their job tasks and have to be able to travel to and from their work sites to accomplish their mission. A security company that cannot protect the client's personnel, equipment and sites or cannot get the client to the areas he needs to travel to in order to complete his work is useless.

RISK should be an accepted part of the work in high threat areas for anyone who decides to work in a high risk area or war zone. For the PSC supplying PSD teams or static security services risk is multiplied because they have the risk to themselves but also they assume the risk for their clients. THAT is their job which means they will need to work in the threat environment with their client's job needs in mind and be able to perform their task which is to protect the client, so the client can do his job. As a security service provider PSCs need to learn what is an acceptable level of risk in order for everyone to accomplish their mission.

CHAPTER 1
WHY A PSC PROVIDER IS HIRED

The use of a personal security company for protection in a hostile area is a decision that needs to be made at a corporate level. Not only will it be expensive but it can drastically change the perception of the company in the area in which it wants to operate. It can also either increase or decrease the risk to the company personnel (clients) that have to live and work in the area. The decision to employ a PSC is never taken lightly or without some thought, and the company usually already has objectives in mind that it wants the PSC to achieve.

Remember when someone hires a security company their profile will go up in the area even if it is a low profile company. They will have new people working, new vehicles in the area of operation, new rules and restrictions, etc. But when a PSC is necessary to work in some environments the methods that you chose to counter the threat to the company that has hired you will determine if you are protecting their business interests. In hostile fire areas, war zones, and other high risk or crisis situations, especially in those areas of the world where kidnapping, crime in general and other types of attacks against foreigners or foreign companies is rampant, more and more government agencies and companies and private enterprises are going to the PSC to help them operate effectively and allow them to continue their mission.

There are things that a security provider needs to do for the client and things that need to be accomplished during your time working together for you as the security provider to meet the client's objectives. There are also certain things that will have to be provided to the client and things the security detail people need to possess in order to do their job. It is extremely important that anyone providing security needs to have training and have experience. Standing guard at a warehouse in North Dakota does not give you the experience or training to run a static guard force in the middle of Iraq. Personal Security Detail personnel need to be thoroughly trained in the jobs they are going to perform or oversee to allow them to do their job in a professional manner, and with confidence. There are so many things that can go wrong or that can happen when transporting someone to a mission site, waiting there for him to complete his mission and then transporting him back to his safe area that saying your experience is "I drove for the general in Germany for two years" just doesn't cut it when the shit hits the fan.

CHAPTER 2
PROVIDING PROTECTION

There are many types of security that can be performed but the two types I will spend the most time talking about are personnel security and physical security of facilities, property and equipment. The main mission of any security detail will be to provide a secure working environment that allows the clients to concentrate their efforts on accomplishing their mission or the task at hand.

So for a PSD or static guard force the main objectives are: avoid any threats to the client, detect any threats to the client and, when detected, counter any threats to the client. This will allow a PSD to defeat the potential threat (criminal, terrorist, insurgent, etc.) with intelligent planning, unpredictability, and avoidance of danger, and also to heighten personal safety and security while minimizing the intrusiveness of security which can make it impossible for the client to do their job. This requires a balance of protection, which is a way of saying that there will always be competing concerns between the client and the security provider, i.e. the need for certain levels of security versus the need of the client to travel, work outside or interact with others. Examples: engineers that need to move to sites, contracting representatives going to meet their clients, giving tours to local politicians, crews that need to work outside exposed, such as movie stars, celebrities, medical personnel. These are all commonly encountered when working security details.

PROTECTION STRATEGIES

1. A PRO-ACTIVE PROTECTION STRATEGY is probably the best bet when it comes to the type of overall protection that can be provided. Pro-active protection is more of a prevention based method that involves multiple tasks taking place continuously to ensure the client's safety. These tasks include but are not limited to:

- risk reduction techniques such as surveillance detection teams and fixed point counter surveillance

- running your own source operations in the communities around your living areas and your other primary areas to keep track of new people, unusual activity, etc.

- risk assessments

- analysis of threat levels

- lots of interaction with local police, intelligence and counter-terrorism officials.

It also includes all of your preparation for movement or for receiving people at a fixed site, advance work at each site to be visited, route surveys, anticipating what is going to happen, resource allocation, security and safety plan development. (You can't absolutely prevent everything, but you can certainly try.)

2. A REACTIVE PROTECTION STRATEGY is a response (reacting) to an act that is taking place or has already taken place in the last few seconds or minutes. This type of response usually only has two components:

- First is the **evacuation of all non-security personnel**, i.e., the client or clients. Get him out of the area as quickly as possible and too one of the designated safe areas for his protection.

- Second is **neutralization of the threat**. This usually happens when you do not have the ability to evacuate the client immediately and must stay in that general area. Neutralizing the threat ensures your client's protection until he can be evacuated to that safe area.

PERSONAL SECURITY for the client is the responsibility of those members of the protective team who are physically with the client whenever he is away from his secure area, including movement between sites and providing security once they get to those sites or work locations that they are visiting on a short term basis.

When talking about **PHYSICAL SECURITY** we are talking about the advance teams that will secure the sites being visited and are responsible for the other duties associated with advance work. We are also talking about the static guard force, any physical barriers such as fences, doors and associated hardware, personal protective equipment, such as body armor, protective helmets and protective masks, and armored vehicles. Both of these areas require planning to accomplish, which entails many subtasks such as intelligence, long range planning, route reconnaissance, threat assessments, physical security assessments, etc. I will cover these subtasks later in the book.

In a standard situation the client's time is generally split between the office and the residence, with minor detours and changes throughout the day or week to other primary areas. A client working in a high threat area usually lives and does administrative work in one secure location, but must travel out to work sites and other locations for meetings with other companies, governments or their clients. There is no way to gauge how much time a client will spend at his secure location, other locations he needs to visit and the actual work and/or construction sites. In the normal environment such as a client's home country, a low or medium risk area or if you are protecting an individual (instead of having a company as a client and being responsible for many personnel at once) the normal VIP is usually at his primary residence around 40% of the time, and at the office or other primary locations about 60% of the time.

The most predictable time of movement for people, regardless of where you are located, is in the morning hours. In high risk or high threat areas the military or government usually restricts movement along controlled routes anyway, but there and in any other location this is the time it is hardest to change a routine or alter a schedule. But during the evening hours, the client is much less predictable because there is nothing set in stone, such as the time they depart the office or stops at other primary areas on the way home.

However, if you are working in that high threat environment where movement is controlled by the government or military then your evening movement can also be very predictable, maybe not the times but the routes used. This is very important because when you and/or your client are Time and Place predictable then you are a target. Predictability is the biggest enemy when protecting a client. As a member of a security detail you need to keep your client from being time and place predictable as much as possible, without interfering in your client's business.

ROLES OF A PROTECTION SPECIALIST

The close protection provider, static security provider, or personal security detail personnel can have many roles when working with the client that are not directly security but all are security related. Their roles include the following:

- **Planner, coordinator, liaison**

- **Facilitator** (make things happen, fix it, make it work)

- Selling yourself as a **service provider** through additional conveniences

- **Protection**. You usually only get clients who have a threat against their life, or live and work in a high threat or high risk area. So during travel you will need to have someone there ahead of time, register early, check elevators and stairwells, go into the rooms to check sinks, check toilets, check the lights, turn on the television, go underneath the beds, and be aware of other personnel on the floor as well (times the maid service works and hours for room service).

- **Escape artist**. Always be aware of the nearest escape route, not only for security reasons but for safety reasons. For example never take a hotel room above the seventh floor because most fire trucks cannot extend their ladders past that (in some places however they go to the tenth floor).

- **Fighter.** You must have the ability to fight to protect your principal, which is your last option as a protection specialist, evacuation to another secure area being your primary goal.

The are three standard things you as a security provider will be trying protect the client from: assassination, kidnapping, and injury. Depending on the celebrity of your client you may also try to protect them from any embarrassment whether intentional or not.

PRIMARY OBJECTIVES

The primary objectives of a full time security contingent, including both PSD and static security, are to deter anyone who has a hostile intent towards your client, detect any threats against the client through the use of intelligence and counter-surveillance, defend the clients when attacked, keeping them safe so you can do the most important thing for them which is to evacuate your client out of the area, get them off the "X", out of the kill zone and to a safe area. You are not there to stay in place and fight a battle. You only engage as long as necessary to EVACUATE your client or clients. You are also not there too bully or threaten media, beat up people, or take other actions your client wants toward people that are not a REAL PHYSICAL THREAT. Be professional.

FUNDAMENTALS

The are five necessary fundamentals for the protective services provider to adhere to, regardless of whether he is providing services for movement or for static sites:

- Prior **planning** before the operation begins

- Proper **assignment of responsibilities** and ensuring everyone knows what their responsibilities are.

- Having the availability of **resources** to do the job. This includes personnel and equipment.

- **Control of information**. This is necessary to contain anyone attempting to gather intelligence on what you are planning and the plans of your client.

- **Flexibility** to change and adapt to any information, situation or circumstance as it arises to keep your client safe and **allow him to complete his mission as you accomplish yours, which is keeping him safe**.

AREAS OF PROTECTION

There are **THREE MAIN AREAS OF PROTECTION** we will discuss when talking about PSDs being used for movement and protection while a client is away from his safe areas.

The **INNER PROTECTION AREA** is where the protective detail is the closest to the client or clients.

The **MID PROTECTION AREA** is when you have plain-clothes personnel conducting surveillance detection and keeping an eye out for any overt or covert threats that cannot be seen by the inner protection detail for various reasons.

The **OUTER PROTECTION AREA** is that area where you have outside agencies such as uniformed police, and or uniformed guards or protective detail members from your company. Using outside agencies requires good pre-event coordination but these people in the outer area are used as a visual deterrent. The outer area could also include counter sniper teams and other reaction forces as deemed necessary during the risk assessment.

THREAT LEVELS

There are also 3 levels of threat, High, Medium and Low, that you will be considering in assessing what security package or arrangement will be best to meet your client's needs.

The **HIGH THREAT LEVEL** is used when the threat is high because of the location you are currently operating in, the areas you will be moving to, what your client does, who your client is, who he will be associating with or meeting and/or any threat specific to your client.

The **MEDIUM THREAT LEVEL** is used when there is no specific threat against your client but a large general threat due to activities taking place in the operating environment, such as random attacks on foreigners, IEDs, etc.

A **LOW THREAT LEVEL** is when there are no specific threats against the client and there is little or no violent activity taking place in the operating area. Remember you must always be aware of your environment/surroundings. Situational awareness is a necessity regardless of the threat level and must always be practiced since you never know how much time you will have to react until it is too late and the attack has occurred or you have missed the opportunity to bypass the threat. Whenever you are working, you must always observe your surrounding environment, evaluate everything, normal, not-normal, threat, non-threat, etc. and be ready to react to any incident, going over your plans of actions and continually scanning the area, observing, evaluating and reacting.

Providing protection for its personnel is a big responsibility for a company and they pass this responsibility on to security providers. It's hard for many security personnel to come to understand that they are not a military unit any longer but a private security contractor. When hiring a security provider you need to make sure they understand their job which is to protect the client. Everything they do comes down to this one thing. They do not engage in sustained fire fights. They do no leave the client to recover other members of their security detail. They do not leave clients to assault an attacker's position. Why do I mention this? Because this has all happened in Iraq and Afghanistan, where security companies have left clients in kill zones, or outside secure areas in order to get into the fight. This cannot be a policy for any private security contractor.

20

CHAPTER 3
THE CLIENT

WHO IS THE CLIENT?

Before you can even start advance work, you must know who the person or company is and, in the case of an individual, what rank or position they hold in the company hierarchy and the equivalent in local and host nation standards.

Company Background: For a company, a background on the company can tell you what other names it has had, does it sponsor sports teams, etc.

Individual Background: If you are protecting one individual, a short biography on them will help the advance personnel know how the client got to his or her position, schools attended, etc. This can assist the advance in building rapport when they get to the site, because often site security personnel and staff know little or nothing about the client or clients coming to visit.

By knowing a little about the client, site personnel or the people who are the subject of the meeting can better relate to the client, and help humanize him/her so they are not nervous, angry or upset when the client arrives. Also having this information available for personnel at the site allows them to become slightly familiar with the client and this could prevent an embarrassing situation from arising for everyone.

The advance should find out as much as possible about the site to be visited and its personnel. Knowing the history of a facility or a city, or knowing people, greeting them by name and knowing small things about them and/or their family and business will greatly help the client develop rapport for his job.

The PSC, and especially the detail leader, should attempt to get the client's likes or dislikes. When dealing with a large company where you may have 30 or 40 people you will transport during a week or month this may get difficult but keeping a short file on each person and briefing the security detail before each movement will be very helpful in avoiding the client's dislikes. This in turn will help prevent the clients from getting upset with the security detail. Remember most of this type of security work goes up for bid every year. If the client is pissed or upset about things the security company does for him or his company, it will be that much harder to renew the contract. Also, knowing little things about clients can help put them at ease during movements.

Try to determine the individual client's opinion of security for the company and in particular their opinion of the security detail and the static security if your PSC is providing that also. Some clients do not mind having a visible security presence around them and understand the need for it, while other clients will dislike a large visible security presence especially if they think it is not necessary, is a hassle to deal with, or is more trouble than it is worth. Getting to know your client's general overall opinion of security and of the security being provided that directly affects him or her will help determine how security is placed at the site and will help prevent the hiring company or the clients you deal directly with from getting upset with security.

The security detail should also know about any health concerns for the client or clients both during movement and at static locations. They should try to obtain, if at all possible, information on any health concerns of the client/clients. This could be very important because the security detail and the advance may have to acquire additional equipment to meet these needs such as a way to keep medications cold, or the location of hospitals or clinics that will have the personnel necessary to deal with that medical condition if it becomes necessary.

All members of a personal security company should have advanced first aid and life saving training which includes advanced CPR Training and Combat Life Saver Training. As a company you should do annual training in this and could even offer this training to the clients if they are interested.

If a client has a physical handicap it may require the detail, especially the advance, to find alternate arrival and departure points, such as handicapped entrances and exits for wheelchair bound persons. Some foreign countries and locations do not have these so a ramp might have to be made or personnel assigned to carry the client up stairs to gain access to the facility. The advance could also have an extra chair set up and ready to go as the client arrives. This will prevent the delay caused by having to pull one from the vehicle and set it up. If the client is hearing impaired, extra batteries should be carried.

You should also have the client's blood type. This is very important in case of an emergency because hospitals in the area may or may not have the blood in stock. It will allow you to find which ones do, and give the hospital notice so they can obtain some, and have it ready in stock for long stays in case of an emergency. Knowing what will be necessary greatly enhances the advance's ability to plan for it and make things go smoothly.

ETHICAL CONSIDERATIONS

Why does this person need protection? This is one of the questions you need to ask before you begin. If you don't know beforehand, you could find yourself with some serious ethical issues to work with or try to work around. You could have a client who is working in a high threat areas or a high risk areas, someone who is famous or a major company executive, or it could be someone who owns a bunch of sweat shops, a drug runner or someone who traffics in people for prostitution. There could be gang problems or it might be a current dictator of a bad country or a former dictator needing protection.

What if its the client's personal life that may give you problems? What if he takes drugs, is an alcoholic, and beats his spouse or children? Likes the night life, hanging out in strip clubs or other types of bars? Asks you to set up plans so he can go somewhere and cheat on his spouse? What if he or she likes young girls or boys? Wants you to deal with paparazzi and other media? Is always getting into fights? Wants to carry a weapon? Wants you to deal with his personal life, throwing out girlfriends or boyfriends?

There are going to be many things you may have to deal with when doing security for others and you must make sure your personal feelings or lifestyle objections do not get in the way of doing your job, which is protecting the client to the best of your abilities. If you start questioning yourself and your client's lifestyle it's probably time to find another client or get into another line of work.

CHAPTER 4
COMPONENTS OF PROTECTION

Regardless of how you feel you should always be professional in your conduct and mannerisms. Be polite and be humble whether you are in a personal security detail, working static security or doing close protection/bodyguard work. When you are doing close protection or bodyguard work you will be going with your client into restricted access areas such as business meetings or luncheons. These are small gatherings so how you look and act will reflect on your client and how he is perceived by his business and social associates. Your appearance and conduct will be very important even though you are in the background with no voice in the proceedings. To prepare yourself beforehand for close protection work you should know where you are going and what you are doing, and wear the appropriate attire to blend in to the working environment.

You will need to have the proper mindset, by conditioning yourself mentally as well as physically to protect your client regardless of the working environment, to be very diverse in your thinking, tactics and techniques, to always be flexible and adaptable, and to be disciplined. But never ignore your instincts. After all this is your subconscious side and it may have noticed something you didn't because of distractions in the operating environment. Your instincts are your survival side of your situational awareness. You will need to work on cultivating your instinct and must learn to trust it and not second guess yourself. Having that sixth sense will help you in most things you do, so trust yourself.

Most of what we have talked about so far is all pro-active work that is done before your client ever leaves his safe area. Now we are going to get into the process of actually providing physical protection, at locations for the client, during movement and its various components. The first thing we need to remember is protecting a client is always defensive in operations and tactics.

THE BODYGUARD (BG)

The bodyguard can work two ways, in a stand alone role or as part of the protection detail. Bodyguard is a term I like to use for the person who has the primary responsibility for the client. He is the one who is always with the client, constantly by his side. The bodyguard's only concern is to protect/evacuate the client. He will follow the direction of the security detail during incidents but he is the guy who grabs the client and takes him where he needs to go.

THE BODYGUARD AS PART OF A PROTECTIVE DETAIL

The bodyguard will fall under the command of the protective detail leader. The bodyguard's only responsibility is the immediate protection of the client. The protective team's role is outer protection, transportation, advance work, and when necessary to help the bodyguard with extracting the client if an incident occurs.

There are certain things that the designated bodyguard should know before each trip with his client to better help him in his duties and to help him function more efficiently with the protective detail. He should know the route to each location that will be visited that day. This means he will know when they are approaching choke points, where safe havens are located and alternate routes if necessary. He will know if the driver takes a wrong turn and in case of a driver down incident in his vehicle he knows where he needs to go with the client.

The BG must have detailed knowledge of each location to be visited. It is best if the BG gets the chance to visit these sites in person but if not then he needs to read the site surveys, trip reports and view any photos and videos taken by the advance. It is important for the BG to know as much as possible about these locations so he can guide the client and so that he will know where to go if there is an incident and the detail team leader gives him directions. Part of knowing about the site locations will be knowing where he and the client are going to debus and embus since departures and arrivals are dangerous times for the client. Knowing where this will take place and the site layout will give the BG the information he needs to find a safe area if anything goes wrong.

The BG will also need to know everything the protective detail is doing and he should have a great working knowledge of their SOPs. He should know all the members by sight if possible and he should know of any other security measures that are going to be taking place or are already in place. This prior knowledge of other security will help everything run smoother and prevent any misunderstandings. Finally the BG must have communications with the protective detail with a compatible personal radio system or cell phone. This will allow the protective detail to pass on information and assist the BG in the performance of his duties. Always do a communications check before leaving for the mission. Ideally the BG will be a member of that security detail, whose primary mission is BG duties.

The BG has to be at his highest level of alertness, along with the protective detail, during the arrival and departure sequence. Most facilities have a location to drop off and pick up people, even VIPs. Since this is a fixed location, it will be known to any possible threats. There should always be some rehearsal/practice on getting into and out of the vehicle. Perfect practice will make it perfect.

During the event the BG will be doing his primary job, that is escorting the client where ever he goes, remaining near but in the background, allowing the client to do his job. There may be times when you cannot do this: meetings you cannot attend, other people's security not allowing you in, the event host's security keeping areas blocked off to participants only (prior coordination is very important), and business deals the details of which might be off limits to anyone not working directly for the employer. If this happens, don't panic and don't bully. You have no legal standing. Be polite, be diplomatic, be FLEXIBLE AND ADAPTABLE to any situation, and don't forget to brief your client on what to do if an incident occurs and you are separated, and that would be to FIND YOU.

Communication between the BG and the protection detail should flow both ways, with the BG sending updates to the detail leader about location and events taking place. The communication needs to be two ways to keep everyone involved with the operation informed of what is taking place and most importantly the locations of the clients. That way the team will know where to look if an incident takes place and will be prepared when the client is moving or leaving.

The final duty of the bodyguard is the post operation after action review and hand over to the next security team that will be responsible for the client. This will usually be a static team at a fixed site, but it could be another BG and protective team. If this is the case this new team needs to be briefed up on everything concerning the client. The BG and the detail leader need to do this together. First they will go over the operation start to finish for any changes that took place to the original plan, including the timing, and any problem areas. Allow the other members of the protective detail to put in their observations, then put everything into a clear report format that is filed for future reference and use. Then the detail team leader and the BG need to get ready for the next operation to take place: what type of clothing is to be worn in the operating environment, has a survey been done, route analysis, assessments and surveys, operation start time, etc. This type of work is never ending. You are always planning and preparing for the next operation.

THE STAND ALONE BODYGUARD

A bodyguard may also work by him- or herself. This is not the best situation for a BG to be in since he will be working on his own without the support of a protection detail. This can be either be a very non-threatening assignment or a high threat assignment depending on the operating environment. In high threat areas it is not the recommended way to use a bodyguard. The major factor in a BG operating alone is no immediate support and dependence on outside security forces for incident support. This places a much greater responsibility on the BG and is a very demanding role.

Everything we have already talked about still remains in effect for the BG working alone. He must have intelligence and knowledge of the sites the client will visit and he must have some type of communications to a supporting agency even if it is to local law enforcement. So the BG will need to gather all the information possible for the planned movement/operation, and conduct risk management which includes a threat assessment and a vulnerability assessment, to the best of his abilities. Then he will need to plan the operation, which should be uncomplicated and easy to remember.

The individual bodyguard uses his common sense, skills and knowledge to keep his client safe. Get as much info on the routes, and the pick up drop off areas of the event and or operation. Find out what other security will be there, then brief the client separately and finally brief support staff, which at a minimum

should be a driver.

During the event the BG must be in a mindset that matches the operating environment. Since the BG is on his own in protecting his client he might need to take fast aggressive action in the event of an incident by removing his client from the threat towards the nearest safe area. This could be exiting a room, forcing the client to the floor behind cover, pushing him into a closet, and it might even include engaging the threat to protect the client.

The BG also needs to be prepared for changes in schedules, arrivals, dinner, etc. This happens all the time and is outside the bodyguard's area of control. Learning to handle changes and remain diplomatic and polite can be a very hard lesson to learn, especially in stressful situations. The final step is an after action review of your actions and events of the day, including what went well, what could have been done better and what to prepare for your next operation.

How a bodyguard dresses and acts is very important to the client. Unlike a full personal security detail, the client will be in situations, meetings and events where people with exposed weapons, full battle armor and assault vests do not fit into and will not be conducive to the business being conducted. Since a BG's conduct, bearing and dress can be important factors for a client, it is important that you know the plans of your client and the types of events to which you will be escorting him. You should dress in a conservative manner. You do not need to dress exactly like the client or attempt to be his clone, but a nice conservative approach will fit into most situations. Avoid light suits and very dark shirts, and anything multi-colored. Remember you are not a fashion icon, so avoid drawing attention to yourself.

You are there to protect the client so make sure your attire is also functional for anything you may have to do. You don't want to rip the seat of your pants out if you have to bend over. You need to actually try on your clothes with the equipment you need to carry and make sure you can move and function properly including bending over, jogging, getting on the floor, getting up, etc. Test your gear and clothing BEFORE you go to the event. You also need to ensure you will not be flashing a firearm or radio gear at people, so have someone from the protective detail or other security personnel watch you go through the motions and tell you if this is happening. Also, always take extra clothing in case of an accident like spilling a drink.

Take your lead in dress from the client because some of the events he must attend or things he will do will not all be in business attire. He might be shopping, going out to inspect a facility or work site, or socializing with family. Use your common sense and do your best.

A bodyguard is not used or needed all the time and in most high threat or high risk situations, usually a personal security detachment will be the protection of choice for moving and ensuring the security of a client or clients, but knowing and understanding the bodyguard's function is important to understanding personal security work.

ADVANCE WORK

Prior to any operation or movement you will need to get as much information as possible about the routes, locations to be visited and personalities involved. This requires you to go out and do a recon or site survey of these things, which is also called doing an advance. This is a major part of security work and will help with planning the operation and making everything work smoothly.

So what is advance work? Advance work occurs before you move the client, and is all inclusive of those activities useful in gathering as much information as possible about a trip or movement and the facilities to be visited. It is done in an effort to eliminate surprises and to prepare for potential problems, incidents, or emergencies when you are away from your client's normal secure areas. The advance is part of risk management and needs to be done to help complete the risk assessment process, by looking for threats and vulnerabilities at each location your client will visit and along each route you may take to get to these locations. Once you have worked out your risk assessment then you can plan for how you mitigate the threat and allocate resources to eliminate or reduce those risks to your client.

The advance will also give you and your security people a sense of the physical layout at each of these facilities, routes and other locations. A prime example of this is when moving to an airport to pick up clients you have to know the layout of the airport, develop contacts with airport security and the customs branch, know where the VIP Lounges are located and how to gain access, what immigration issues you might encounter such as expired visas and or bringing in electronics that have to be regulated and how to deal with those problems while at the airport, as well as lost luggage contacts with each airline so you will know who to contact and how to contact them to recover it. Or when staying overnight away from your normal secure areas, you'll need to know what hotels have the best facilities to meet the needs of the client and your needs in protecting him, how the hotel rooms are set up, the location of all emergency exits and where they lead to, the availability and functionality of fire alarms and fire fighting equipment, the emergency services response time, what is the tallest ladder truck they have, what floor will it reach, etc.

As for routes of movement you need to drive them at all times of the day and in all types of weather conditions to get a sense of time and distance between the stops on your trip.

An advance also gives you the opportunity to establish contact with people who can help facilitate the movements to these locations and any overnight stays. These people may also be able to help you in an emergency situation or if a non-emergency problem comes up that you and your people cannot deal with yourselves. When you are making these contacts, do not leave out people like the service personnel, door man, bell captain, head maid, security personnel, and valet. In some locations they are the ones who can find ways to help without raising the profile of you and your client, so use your rapport building skills to develop and maintain these relationships.

You should also be sure to make contact with the local police and military whenever you can, because knowing people in positions of power and cultivating those relationships could have unexpected benefits some day. If necessary do not be worried about helping them whenever possible, even if at that time there is no direct benefit to you. I am not saying bribe someone but give and take is an accepted way of doing business and cultivating friendships in most parts of the world. Do not forget to contact the on-site security and other facility personnel at each location, including other PSDs that will be traveling to the location with other clients. It is also a good idea for the security leader to get to know the client's administrative personnel or those who decide what facilities they need to visit and make their traveling plans and accommodations.

Advance work can be very simple such as a few phone calls and a small advance team to confirm information or it can be a full blown advance that could take a full advance team several days to complete. This will all depend on whether you have been to the location before, the type of operating environment and the threat. An advance team should be done by members of a security detail, if possible personnel who have contacts in this area or who have done a mission to the same area or have knowledge of the area, route and location. The work done in the advance depends on many factors, such as risk assessment, familiarity with the destination and comfort with the local contacts.

What will a full and complete advance entail? At least one member of the advance should be a member of the client's primary protection detail, who will be on the preliminary trip to view the local area and get to know faces. I prefer sending a driver from the primary detail so he can drive all of the possible routes and get a feel for the area so when he returns with the client he will not need to get directions from the navigator and knows where he will need to go to debus and embus, park, get gas, etc. The advance detail will also provide recommendations as to routes of movement to all primary areas and facilities to be visited with alternatives. Alternative routes are very important and need to be looked at with careful thought.

The advance will also make all the necessary arrangements for luggage pickup and drop off, vehicle staging and parking, and they will confirm reservations and reserve rooms on several different floors for the client and security detail. Then they will pick a room at random to stay in for the mission. This may mean you have to pay a fee when you decline the other rooms but it is a small price to pay for not allowing anyone to know beforehand what room your client will be staying in. Also I always get one room in the name of an advance member on the third or second floor. This member will not be associated with the detail. I like to use this room as a safe haven in case of an incident at the hotel.

Then the advance will visit each of the facilities and or venues to be visited by the client. They will once again make contact to determine debus and embus locations, vehicle staging, static security currently in place, communications plan for security, etc.

All of this information will be compiled in the advance report or site survey to allow the protective detail leader to have a good working knowledge of the area, and photos of contacts and routes. If it is done correctly then the protective detail leader can simply follow their plan of action. I always encourage the use of photos and video in the report to give the protective detail members an initial view of the operating area.

When collecting your information about the routes and different facilities you need to ensure that you record the time and distance between points. You also need to include on long trips the point of no return, which is the point that once you cross it you have to continue on to the planned destination or refueling area and will not have the ability or support to turn around and return to your point of origin. The names of the contacts at each point should also be included along with a photo and phone number. Whenever possible a member of the advance should accompany the security detail on the trip, along with the protective detail member who accompanied the advance this will give you two personnel who have made the trip and have first hand experience.

When the client arrives at each planned location, a small advance team will move ahead to the next location to ensure there are no problems. The advance should be moving low profile and meeting their contacts and arranging for arrival security. They will also make sure there are no problems along the routes or at the next destination and they should always maintain constant contact with the security detail. On long trips the small advance moving in front of the detail can also be used as a quick reaction force in case of an incident.

When should you have the advance work done? If it is a facility or location you will visit a lot, an advance should be done and kept on file, with a trip report. Then each time you go to that location again one should be completed very near the time of the movement. It could be a few hours ahead, or several weeks ahead depending on the length and complexity of the of the operation. When you are conducting a simple movement and operation into an area that you have knowledge of, with roads and contacts you are familiar with, then the advance should be done close to the departure.

If it is a complex operation that has a high risk with lots of new territory you should always ensure an advance is done at least 2 days prior to the trip. It is important that after each trip you do a trip report, in a format that lists who went, what routes you took, what the trip was for, gas used, time it took, distance, any observations you had and any photos or videos taken. This will help ensure that over a long period of time you do not continually use the same routes, same gates, same stop over points. By reviewing the old trip reports and changing these for the next run, you avoid getting into a routine and being predictable. Being predictable will make you an easy target.

A proper advance will reduce the risk and facilitate movement of the detail. It also eliminates any surprises and provides contingency plans in the event that the original plan goes wrong or an incident causes it to change. You should never move to a new site or RON in a location you have not done an advance.

THE PERSONAL SECURITY DETAIL (PSD)

PSDs are used primarily to move clients around high risk areas and war zones, and to secure facilities during temporary visits and provide physical security for fixed site temporary client locations. To begin with, whatever type of service a personal security company (PSC) is attempting to provide it comes down to one goal, SECURITY to the client. To do this a security company or detail has to achieve a **360 Degree Area of Seamless Protection** for the clients. This security bubble or circle of protection will consist of three layers.

The first is the **inner layer** which is the area immediately around the client. The inner layer can include the use of a bodyguard, and also include a close protection team which is responsible for the immediate protection around the client. This is your primary protection team.

The **second layer**, which encompasses the inner circle of the primary protective team, consists of additional security, possibly meaning law enforcement, or event security staff. This layer will also include walls, fencing, hesko barriers, road blocks and other types of perimeter barriers. It should have some type of emergency medical care, in addition to the trained medic that is part of the primary protection team. This layer will have the surveillance detection team for both facility and personnel protection of the client's assets. Good communication is important with the counter surveillance teams. Since they will be on the outside looking in, hopefully they will see any suspicious activities and pass them immediately to the detail team leader. The surveillance detection team members should not take any part in the actual security around the client and should not interact with the visible security teams. You do not want them identified in the event that a potential threat is also watching the event.

The **third layer** will be the security measures that take place when an incident occurs, including the counter-assault team, intelligence assistance, emergency evacuation measures for the client, and fixed security elements. They can use technical observation methods and recording devices including closed-circuit television, use of sensors, seismic, infra-red, or face recognition software. They will be working out of the security center or security office passing on information as needed to the detail leader, doing mostly observation support until an incident occurs. At that point, they will take directions from the detail leader or team leader and perform those functions he requires to keep the client safe.

EQUIPMENT CONCEALMENT PROBLEMS

When working, depending on the type of clothing you will be required to wear, the concealment of weapons, and communications equipment might be necessary. This can be a more difficult problem for women. Whereas men normally wear belts and jackets, women, depending on the event, usually don't. They are in dresses or slacks which don't require that they wear belts. The other problem will be the means of communication used because women normally don't wear articles of clothing that have the same amount of pockets as men. But women have other options, such as the type of bag they can carry, which can allow them to have a weapon available.

Ankle holsters can be very good in cars but require lots of practice to put into use. Regardless of the weapon you are using and how it is concealed you must practice, practice, practice. During the operation you must at all times be totally aware of your weapon and how to draw it and put it into use depending on your operating environment. This has the potential to be critically important. Also, practice putting your weapon into use while you are running, walking, going up stairs, going down stairs, sitting and any other activities you may have to do in the performance of your job, because Murphy's law says that you will not be in the best position when it is time for you to draw your weapon and engage a threat. This is assuming you are working low profile. If you are working high profile, high threat then your weapons will be visible anyway.

LOW VISIBILITY VS. HIGH VISIBILITY

Basically there are two types of environments you will have to work in: permissive and non-permissive. With this comes a major decision that will have to be made: should the security be a low profile/low visibility or high profile/high visibility?

Low Profile/Low Visibility: This is when you use up armored innocuous vehicles with no markings at all or set up to look exactly like a local vehicle with tinted windows or window covers. Your weapons are not visible, the personnel who are visible dress to blend into the local environment, and most importantly you obey local laws and make the effort to fit into the operating environment, with no grandstanding, ramming, showing of weapons, etc. If you're good and sometimes lucky, no one will even know that a security team with a client has passed by.

High Profile/High Visibility: This is when you travel and move about like a security company or the military, with high visibility vehicles, SUVs up armored, gun trucks, weapons visible, and security personnel wearing combat vests, body armor and helmets. This is also using aggressive driving tactics, waving people off with weapons, bumping other cars, firing warning shots, and pushing vehicles out of the way. When you are running in a high profile it means everyone knows you are there. They might not know who you are or who your client is but they know when you are coming and when you pass.

This type of profile will get you noticed and the object behind working this way is to present yourself as such a hard target to attack that the threat will leave you alone and go after softer targets with less risk. The one problem with this is if everyone is running high profile, hard target and there are no softer targets available then they will just adjust their tactics and hit that harder target. Case in point is working in Iraq. If the threat is not afraid to engage tactical military vehicles with up armoring and air support, they why would they be afraid to engage a private security contractor?

So this just comes down to opinion, current threat operations in the working environment, company resources and of course what the client wants. In my experience they prefer the high visibility movement and security because someone told them they would be less of a target if they presented themselves with a strong visible presence. Usually this type of profile is presented by security companies that do not have the ability to function in a low profile security bubble either because they lack language capabilities, local information or just do not have the experience available in the employment pool to accomplish this.

THE CLOSE PROTECTION TEAM

The close protection team is the primary protection for the client. Their only mission is to ensure the security and protection of the client. They do this through:

OBSERVATION: The protection detail will be conducting a 360 degree field of continuous observation to provide a bubble of protection during movements through an unsecured area. During vehicle movements each member of the detail has an assigned sector he is responsible for. Observation sector assignments will be based upon where they are located in the vehicle and in the convoy. During foot movement each member of the detail is still assigned a sector of observation, which is once again based upon his location in the movement formation. Each person is looking for suspicious or out of place people or things, and anyone who is keying on the client and correlating their actions to the movement of the client or the protection detail. This can include taking pictures of the detail members, video of the movements and taking notes.

ATTACK RECOGNITION: This is the ability to recognize when an attack is taking place versus some other type of non-threatening gesture or event. It is important to be able to tell the difference because as a security detail you are going to react differently to a non threatening event. If you overreact, innocents could get hurt, your client embarrassed, and possible legal action against both the security company and the client could result.

REACTING PROPERLY TO THE THREAT:
This comes with practice, rehearsals and experience. Using the proper response to negate any threat will be important and will depend largely on the security detail member's perception of what is going on. Training and practice will help you develop the skill to quickly judge the situation and react accordingly to protect the client. Your response to an incident should be proportional to the perceived threat, not an immediate escalation to the next level. Remember your function is to delay the threat just long enough to cover and evacuate the principal.

COVER THE CLIENT: (Protect the client from the threat's attack.) When an incident occurs the bodyguard, if there is one, will step between the immediate threat and the client. If there is no bodyguard then the detail leader will do this. The detail member closest to the threat will move towards it in an attempt to draw its attention away from the client and neutralize it. The rest of the protective detail closes in to provide 360 degree security, and starts moving the client to a safe area away from the threat. Remember to keep your eyes open and don't become focused on the visible incident because it could be a distraction. The advance detail, which should be performing outer security, should move in and attempt to deal with any problems and incidents, since they will be coming in behind the instigator or instigators. This will give them an element of surprise and cause the instigator to turn away from the client allowing the protective detail to access the situation and take appropriate action with the client.

EVACUATE THE CLIENT: (Get the client away from the threat to a safe area.) Once there is a confirmed physical threat, the detail will evacuate the client to a safe area where he can be protected until the incident has passed. The detail might have to get physical to move the client to this safe area by pushing and forcing people out of the way. Once in a secure area the decision will be made whether to withdraw from the site or continue. This decision process must include the client's input! Now the hard part about evacuating the client isn't the actual movement but getting security personnel to show restraint. In most cases you will have no legal authority or right to push, punch, or threaten with weapons ANYONE, let alone innocent bystanders. You are a civilian so you need to use the appropriate response for the situation. The bodyguard or detail leader is the person responsible for the safe evacuation of the client. A successful evacuation depends upon anticipating types of attacks and rehearsing evacuations out of the "kill zone", which means everyone needs to know where the quickest evacuation routes for each area visited are, and practice/rehearse as a team the drills to be used.

NOTES　　　　　　　　　　　　　　_NOTES_

CHAPTER 5
COMPOSITION OF THE SECURITY DETAIL
AND THEIR RESPONSIBILITIES

The security detail can be made up any number of ways. Most companies have their own version, but all have the same basic characteristics and members. Sometimes they will be called different things. There are basically three people and five groups that can make up a security element for a personal security detail and these can sorted by their duty position. The same is also true for any static guard positions that are used by a PSC to provide a full security package, so I will use generic titles for job positions.

The **DETAIL LEADER** is responsible for all aspects of the mission. Sometimes called the Agent in Charge (AIC), he is the first primary leader of the detail and is in overall charge of the detail, including mission planning and assignment of duties and responsibilities for the detail. The detail leader's demeanor will set the tone for the entire detail and mission. In the absence of a designated bodyguard his main function once the mission is underway is the protection and control of the client. He must stay with his client at all times so he is present for all moves. He is the primary body cover for the client in the absence of a designated bodyguard, and should be the only one performing courtesy functions such as opening doors. He is normally in the same vehicle as the client in the front passenger seat during any vehicle movement and is usually positioned to the right rear of the client for foot movement and does not take a spot in any foot movement formations around the client

The **PERSONAL SECURITY OFFICER**, close protection officer or bodyguard is responsible for the close-in security of the principal and accompanies the client at all times when away from a secure area. In the absence of a designated bodyguard the detail leader takes this responsibility.

The **SHIFT LEADER** is the second primary leader of the detail and his duties should include but are not limited to working directly for the detail leader. He should be positioned in the passenger seat in the follow vehicle in the motorcade during vehicle movements. This will give him a view of all vehicles in the motorcade or convoy. He needs this since he is the tactical leader of the detail during the operation. During foot movements he is usually positioned to the right rear of the detail leader and is part of any protection formations.

As the tactical leader he organizes the operation, plans it, then briefs the operations order to the detail after giving a pre-brief to the detail leader. As the tactical leader he gives all security personnel on the mission their individual instructions and guidance as necessary, making changes to the detail's tactics during the mission. He supervises the detail in all aspects and is responsible for any disciplinary problems that may take place during the mission. He is in constant contact with the detail leader and gives him updates on the operation and tactical situation as necessary.

He is also a primary link to the operations center and sets up detail rotation schedules when in a static environment for eating, toilet use, water, sleep, etc. He also is responsible for providing off duty schedules and training schedules for his detail team, when not on missions, including driver's training, tactical SOPs, rehearsals, and range time. The shift leader is also the one making the decision to bring in the Counter Assault Team (CAT). He will decide if this is necessary and when they should be called in to assist.

SECOND IN CHARGE (2IC): The name of the second-in-charge can be misleading because he is actually the third primary leader of the detail behind the detail leader and the shift leader. He works directly for the shift leader. During movement in the motorcade or convoy he is in the passenger seat in the lead vehicle, He will assume the tactical command of the detail when the shift leader is not available. He is also responsible for the routes and movement navigation and for making sure that the speed of the vehicles in the motorcade or convoy are safe. He assumes command in several of the foot movement formations. He makes suggestions in changes of the tactical plan to the shift leader as necessary.

The **CLOSE PROTECTION TEAM** is closest to the client, allowing only authorized personnel to approach the client.

The **SECURITY ROOM TEAM** is in place at the office or home. A residence watch is created with support personnel in order to run operations in the security operations center while the client is traveling, regardless of distance or time away. A traveling security operations center is always set up when a mission away from any of the client's permanent locations is 24 hours or longer. In order to run operations at the visited site, the traveling ops center will maintain 24 hour contact with the main operations center back at the home base. Security personnel working in either ops center will maintain a visitor control log and a daily journal.

A **BAGGAGE TEAM** of designated advance team personnel will maintain accountability for baggage and equipment, ensuring that it is under visual control of a member of the security team at all times until it is delivered to its destination at the visited site, whether it is a hotel, residence or work facility. They will supervise all baggage handling upon arrivals and departures. A log should be kept of the number of bags and equipment boxes, where they go, in what vehicles and what goes to whose room at arrivals. All bags should be clearly marked with a numbering system that will correspond to a check sheet so those people designated to take care of the baggage will know exactly what goes where without interrupting the close protection team or the client with questions.

The **DRIVERS** take their direction from the detail leader of the protective team. He will be the communications point for all personnel wishing to talk with protective team members or to pass on information. During the pre-mission briefs the advance team should have already briefed the drivers on the routes, stops, arrival and departure points. Any changes must come from the detail leader. Drivers are also responsible for the preparation of their vehicles, and in keeping the motorcade's formation and integrity intact from mission departure point to mission end point.

The **ADVANCE** are the individual(s) who are responsible for the planning and establishment of security at the site(s) the client intends to visit. If possible, a detail member conducting the advance should also have a hand in making the activities schedule such as the living and eating arrangements made for the client's and security detail when they visit a new area, city or facility.

The advance's responsibilities include guiding the motorcade into the arrival/debus site via radio and/or telephone communications. If there is no advance team member with the motorcade or if there have been changes, they will give continual updates to the shift leader about the current situation at the site prior to the client drop off. Upon arrival the advance leader will ask the client to follow him to the safe area the advance has set up, which could be an office or the client's room.

The advance will also have guards pre-positioned and posted for the static security of the site during the client's visit. If available, he will employ explosive ordinance disposal (EOD) or bomb dogs to conduct a search. At a minimum each room the client will visit is searched by the advance using facility bomb search techniques.

The advance will prepare the mission site for the arrival and departure of the client and for egress to a safe area if necessary. This can include, as part of the site survey, any possible Helicopter Landing Zones close by and accessible, nearest hospitals or emergency medical facilities, defensible positions located at the site or facility or in structures close by, and all the routes and movement techniques and schedules to and from each of the possible locations just mentioned. The leader of the advance will also keep the shift leader informed of any changes that have occurred since the mission brief or any changes in the site survey overall.

SECURITY ADVANCE PATROL (SAP): This is a vehicle sent out in advance of the main body of the convoy. It is usually 4 personnel: the driver and 3 armed team members. Each member of the security advance has an individual radio plus one for the vehicle or they could use cell phones as their primary means of communication. The mission of a security advance patrol is to travel ahead of the main convoy, and identifying, intercepting or interdicting any potential threats to the main body convoy.

COUNTER ATTACK TEAM (CAT): Sometimes called a counter assault team or a quick reaction force (QRF), this is a vehicle containing at a minimum 4 personnel: the driver and 3 armed team members. There is also an advanced medical trauma pack located with the CAT, and each team member has an individual radio and the vehicle will have one

mounted. This team will take direction from the detail shift leader. The CAT's duties are to respond to an incident in which the main body of the convoy comes under direct attack or ambush. A CAT is not always needed depending upon the situational requirements and the operational environment of that mission, and sometimes the SAP can also be used as a CAT, and called back to help as necessary.

TERMS AND DEFINITIONS

• **SECURITY OR PROTECTIVE DETAIL**: A group of individuals assigned to protect the client. It will include movement by vehicle, public transportation and foot operations.

• **CLIENT**:   The business and all designated assets to include facilities and individuals for which the private security company has taken a contract to protect.

• **SECURITY OPERATIONS CENTER**: A designated area for security that receives all intelligence, information, communications, and/or questions from the detail during movement. The operations center will be in operation 24/7 and monitor all the security details including static, mobile and convoy movements. There should be a permanent operations center at the main operating location of the client and a temporary one at any short term away locations the client is visiting.

• **MAIN BODY**: The vehicle in which the client or clients ride. It is driven by a person who is thoroughly familiar with the entire geographical area and is trained in defensive driving techniques. There may be more than one in a convoy or motorcade so you could have Main Body 1 and Main Body 2.

• **LEAD CAR/ SCOUT VEHICLE**: A security vehicle driven directly in front of the main body which is responsible for protecting the main body from the front. It is also responsible for the overall navigation of the motorcade.

• **FOLLOW CAR/TRAIL VEHICLE/GUNSHIP**: A security vehicle driven directly behind the main body which is responsible for protecting the main body from the rear and the flanks.

• **CAT VEHICLE**: A security vehicle driven directly behind the follow car which is responsible for protecting the main body from the flank at halts. It is also responsible for quickly interdicting a threat by placing itself between the threat and the clients in the main body vehicles.

• **MAIN BODY DRIVER**: Drives (aggressive or evasive) and conducts counter surveillance.

• **FOLLOW CAR DRIVER**: Drives (aggressive or evasive), maintains integrity of motorcade, uses follow car as shield for main body, and conducts counter surveillance.

• **DETAIL PERSONNEL**: Assist driver in driving defensively by conducting counter surveillance during movement and looking for attack site indicators and pre-attack indicators from personnel and vehicles along the route of movement.

Protective details in the basic mode travel with 2 security vehicles and the main body or vehicle transporting the client. The lead, scout vehicle or the first in line has a driver with 3 security personnel, the main body has the driver and detail leader plus the clients, and the tail vehicle, gunship or last vehicle has a driver and 4 security personnel with the addition of a rear gunner. This configuration allows for a basic 360 sphere of observation, and gives security personnel the ability to deal with threats in their arc of observation without a major shifting of personnel within the security vehicles. (See below for an illustration of this configuration with sight lines for each person.)

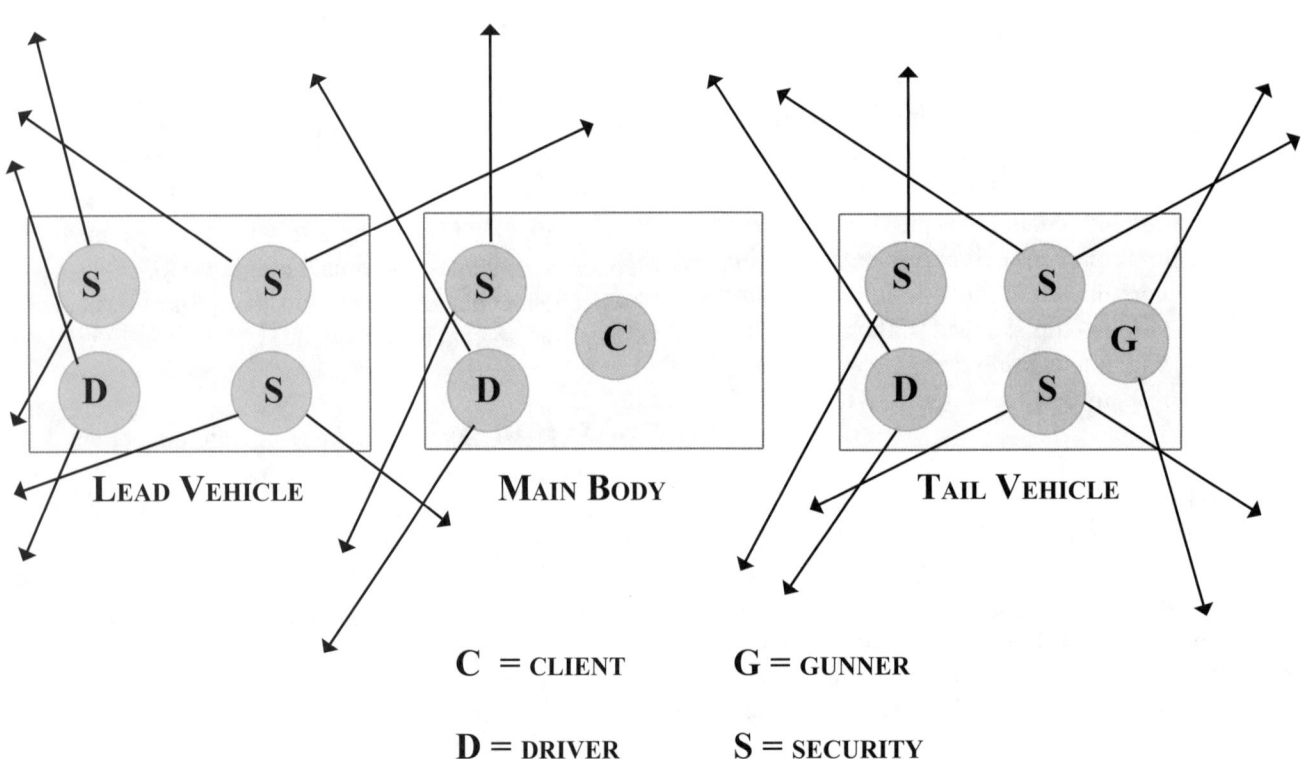

LEAD VEHICLE   **MAIN BODY**   **TAIL VEHICLE**

C = CLIENT   **G** = GUNNER

D = DRIVER   **S** = SECURITY

CHAPTER 6
KNOWING THE THREAT

To get an understanding of the potential threat in the environment where you will be working you need to go back to the one of the most basic military tenets, "know your enemy." Study the area you will be going to. Keep up to date on current affairs for that area, such as local politics, crime reports and even sporting events. Make sure you look at the history of threats in your area of operations. This will give you an overview of how the threat targets its victims, how they plan and how they conduct their operations especially their methods of attack. Then you must follow all new incidents. You need to find out as much as possible about current tactics and techniques the threat is using. Have they changed from their historical patterns? If so this could mean a new source of training, support or resources for them or new leadership. This will make you more aware of their current capabilities, which will help you in planning your security operations. See if it is possible to attend any briefings given by local authorities, law enforcement and friendly embassies. This may be possible depending on how valuable the client you are protecting is.

There will be many sources to help you study up on the current operating environment in your area of operations; one of the best is talking to other security companies working in the area. See if you can exchange after action reports, intelligence from company sources and any videos or pictures taken on runs. Local military, or police might share information with you also, such as tactics and techniques used in the most recent incidents that are not available to the general public. Non-government organizations operating in the area usually have very good local intelligence since they operate directly with the locals they are providing support for. And any local sources you can cultivate are valuable, setting up your own local intelligence net usually pays off in the long run. Finding retired military and/or law enforcement from the country and having them gather information from around your operating environment will give you another source of good information.

Where does the threat to you and your clients come from?

Terrorism: which can be defined as the use of force or the threat of force by individuals or organizations to achieve their political or ideological aim, is one of the primary concerns for people everywhere since you can get caught up in it in so many ways.

Hostile Nation: This is any country where the client or the client's company is not welcome by local nationals, or a country that is having problems maintaining internal security because of strife, regardless of the reason, creating an unstable security environment.

Criminal Threat: In any country this type of threat is always present, and includes organized crime, street thugs, drug cartels, and gang violence. Usually the higher the country's poverty rate the higher the crime rate and the bolder the perpetrators are.

Cranks: These also can happen anywhere and include people who are overly inquisitive into your client or the company, who might think that your client is in love with him or her, and people who hold a personal grudge against the client or company. It also includes street beggars who hassle people for change, winos, drunks, druggies, anyone who might have a chance street encounter with your detail or with the client at his office, a site he is visiting and or one of his other primary areas.

So what does this mean to us as someone working a protection detail, either close protection or fixed site? Most of the time we work in anticipation of events that could happen, this will consume a majority of our time, thinking out scenarios, what-if-ing things that could happen and what we would do to prevent or react. Less than 3 percent of the time will we be actually driving, evacuating, shooting, or functioning in that high pressure situation to keep the client safe. This is the hardest part for most people, living in anticipation, and its why most people cannot do this line of work, but that anticipation, always visualizing, working out scenarios must consume 100 percent of your time while working.

RANGE OF THREATS

The Range of Threats and types of attacks that protective details have to be prepared for include the following:

- Assassination
- Surveillance
- Kidnapping
- Hijackings
- Hostage taking
- Assault on a physical site
- Ambush during movement
- Bombings and bombing threats
- Improvised explosive devices
- Drive by shootings
- Sniper attacks
- Intimidation/threats
- Arson
- Passive and/or active sabotage
- Hoaxes
- Street Crime
- Crowds/Protesters/Rioters
- Random attacks or incidents
- Looting
- Accidental injury
- Illness or medical emergency
- And in some cases embarrassment

There is no specific order of occurrence. Threat levels and what threats you are looking at will change as you move, depending on what part of the country that you are in and what type of environment you are in. Your location, time, and who you are protecting will determine the order of priority for the threat. Don't get locked onto one thing, remember to cover the whole spectrum.

So how do we deal with prospective threats? We have to plan for and conduct good risk management. I like to do my planning by looking at both Internal considerations and External considerations.

INTERNAL CONSIDERATIONS

Internal considerations are those things that you as a security detail will have some input with the client. They fall within your security bubble so you should have the ability to control or alter these things to meet the security profile necessary to provide the best protection or in some cases just do away with things that interfere with your zone of protection for the client.

INTERNAL CONSIDERATIONS INCLUDE:

Client Profile: Does your client have a high profile? Is it unintentional or is there a reason? Is it because of who he is? Who he works for? What he is doing? Something he has done in the past? Where he is traveling to? Statements he has made? All these are things that you as the security provider can control by having a good information control system set up and only allowing out the generic information that will not harm, or increase the risk to your client. Information proofing your security bubble is a very easy step, but very hard to accomplish when clients need to do press conferences, get reported on in the paper, etc. You will have to work closely with the client and explain to him the need to information proof his life in order to help lower his profile.

Client's Family Profile: Sometimes when protecting an individual you also have to deal with family, so all of the above also has to be looked at for family members who are included in your security bubble.

Company Profile: What has been in the media about the client company? Negative/Positive? Who authorizes press releases? What is the PR goal? Do they put out location information? Date/time information on movement of company personnel or functions? Does the company contact the media to pass on press conference information? Photo opportunities for openings or other publicity events? Who controls the flow of information out of the company to the media? Someone in security should have final say over what is put out and who knows specific time, location, travel, and vehicle type meeting information.

EXTERNAL CONSIDERATIONS

EXTERNAL CONSIDERATIONS INCLUDE:

Local political climate: What is going on locally? Are they against foreigners? Is there talk against the person you are protecting because of his country of origin or the business he represents? American companies? Is there a lot of rhetoric against Americans and American interests to help get ratings during local elections? Are there groups that are targeting outside business interests during any local campaigning to gain political support?

Government: Is the country's government or local governments in danger of being overthrown? Are attempts to discredit the local government, law enforcement? Is there a possible coup attempt in the future? Are political leaders keeping low profiles and traveling with large protection details?

Protests: Are there large protests or demonstrations taking place in the country, either urban or rural? Are these protests taking place along routes you need to use? At facilities you need to visit? At embassies or other government buildings? At other business interests associated with the person you are protecting?

Media coverage: Is it positive or negative for the people you are protecting, the company, the country of origin for that company or its personnel?

Construction: Improvements to roads, buildings, bridges. What is going on that will affect your transport of the client or clients, how long will it go on, is it official? Be sure to check with authorities about all construction projects along routes and near facilities that your clients need to travel on, by or visit.

Local law enforcement/military exercises: Are the local police setting up DUI check points? Other types of check points? Do you have to stop or can you bypass with pre-coordination? Will your liaison with local police and military give you the dates, times and locations so you can adjust your movement times, and routes of travel accordingly to deal with traffic and delays?

Parades: Scheduled parades for local events or holidays - were are they planned, advertised, routes put out, parade permits issued, etc. Knowing when these take place and avoiding their routes and any facilities along those routes can be important

Weather: You need to know how different types of weather are going to affect your routes of movements, not only for your vehicles but the local population also. After all if they can't move, they might end up slowing down traffic or blocking roads that will affect you even if the weather doesn't affect your capabilities.

Sporting events: Sporting events affect so many different things when it comes to movement that it needs to be planned for. It can affect traffic, with roads closed or blocked off, pedestrian traffic, etc. I have been in countries where local sporting events shut down the town, so be sure to research and get with local police and your local intelligence net to find out what the possibility is of sporting events interfering with movement.

These are just some of the things that might need to be considered, but not all. Wherever you are working there are bound to be unique events or situations that you will have to adapt to in order to provide the best protection to the client. BEING FLEXIBLE is a requirement for this job.

NEEDS OF THE THREAT

Before the threat can launch their attack they will need several things to help them plan for it. After all they want a successful operation just like you do.

Their basic needs will be:

- the intelligence, weapons and equipment needed to conduct the attack
- the ability to conduct the attack
- the opportunity to conduct the attack (this will be provided by your mistakes)
- access to the attack site
- the ability to escape after the operation is complete

This brings us into the threat's operations cycle.

Understanding how they plan and what they need to accomplish their mission will give you as the security provider indicators to look for during your proactive phase of security, so you will know when someone is planning something and what you need to avoid.

THREAT OPERATIONS CYCLE:

1. Identify the potential targets: During their initial look at potential targets they are looking for targets that will help them meet their overall objective to include their political agenda and media oriented goals. After all they don't want to turn public sentiment against themselves and they want to make sure the message they send with the attack is one that achieves the group's goals. The target has to also be within the attack capabilities of the group. They don't want to attack anyone who is too strong a target or whose security measures are outside their ability to compensate for. Also they want to minimize collateral damage and attempt to limit the damage to just their target.

2. Initial reconnaissance and surveillance: The threat now begins building target folders on all the potential targets by conducting reconnaissance of all the targets to determine the security postures and awareness levels of all the potential targets. They also conduct surveillance to determine patterns and routines trying to find out which potential target will be the easiest target to hit and still accomplish their goals. This phase is the visible hand of the threat, when they have to leave their comfort zone and start actively collecting intelligence against potential targets. This visible portion of the threat is a dangerous time for them and this could be the easiest time to see the threat acting against you or your clients. Since surveillance personnel are very hard to train and might be in limited supply they will use lesser trained personnel and people who might not be that familiar with the environment so they can cover everyone on the potential target list.

3. The target selection: The final target selection will take place after as much intelligence has been collected on all potential targets as possible. The target list will probably be put into some type of order with the easiest targets first. This allows them the option of hitting more than one when practical. By easiest I mean the softest target, which will be the person or organization that has a weakness that can be used to the threat's advantage. Who picks the person to be targeted? _THE TARGETED person does, by their actions or inactions._ A target has a weak security profile and a low awareness level.

4. Closer, more detailed reconnaissance and surveillance: When they have their target picked, they start a more in-depth intelligence gathering mission to find every weakness and pattern they can exploit. Once they get enough information they will determine:

- How the attack will take place

- What type of attack it will be

- When it will be

Who makes the final decision on this for the threat? _THE TARGET does by his actions during this period of the operations cycle._ Remember even though the surveillants will be better trained, surveillance is the VISIBLE hand of the threat before an incident!

5. Plan the operation based on the intelligence gathered: Now that the threat has the who, how, when, and where, they will begin working on a plan that will meet their needs. This is when they come up with the detailed plan, start assigning jobs and missions and seek specific intelligence to support their goals. During this phase general surveillance will be scaled back and the more professional surveillants will come in to take over, so as not to warn the target, law enforcement or other government security forces.

6. Rehearsals of the attack and dry runs: When they have a plan in place and people have been assigned tasks, they will start rehearsing the tactics and techniques to be used, doing walk throughs, and full blown rehearsals away from the attack site. Depending on how well they can blend in to the environment they may even do walk throughs at the actually attack site so that the attackers will be more familiar with the surroundings of the area and the plan.

7. Execute the operation according to plan: This will be the threat's most vulnerable time. They are getting ready to commit a criminal act and must move to the attack site and set up, getting the attack ready to happen. They must do this without raising suspicion from locals, law enforcement and the target's security elements. This will be difficult. Since intelligence personnel are hard to train and some members of the threat only do this type of work, they will pull all personnel who are not taking part in the attack out of the area. This raises the possibility that people who are normally working at the attack site, along the route somewhere or who are normally present in the area for other reasons will be gone at this time. The flower girl and her cart are missing, the coffee shop is closed, the child selling papers at the stop light is not there. These can all be attack indicators that the security team can pick up on.

8. Escape the attack site after execution of plan: This is a necessary step for most terrorist attacks (not suicide bombers). After all, the threat will want all of their people, especially the trained people, to get away so they can be used again after a successful attack. They also must get all or at least a majority of their people away so they can consider the attack successful. If they get their target but everyone dies, while it may accomplish their goal, it is not good publicity which they need to gain support.

9. Exploit the incident using all available forms of media: This is the whole purpose of the majority of the attacks that will be carried out. This is how they sway public opinion, gain support from the populace, show people that they are a force to be reckoned with, show they are trying to support the people and that they are competent and have the ability to do what needs to be done. Even when its just a revenge or punishment attack such as kneecapping by the IRA in Northern Ireland, publicity is necessary to show people what they do to those who betray their cause or get in their way.

10. Conduct an after action review of the operation: After an attack, whether it is successful or not, they conduct an after action review (AAR) to discuss what happened, how things could be improved and what worked well. This is how the threat learns from their mistakes and learn what works well for them and what things will not work at all. Any lessons learned will probably be used in their next attack and they will keep getting more efficient and professional all the time.

Remember, when learning about and planning for the threat:

The point of providing protection is to keep the client safe and in his comfort zone, not get him into fights, high speed chases, gun battles, ambushes or near misses. If you do your job right 90 percent of the time he will never see or know what you are doing to protect him. *The LESS EXCITING LIFE IS FOR THE CLIENT, THE BETTER IT IS FOR YOU!*

CHAPTER 7
RISK MANAGEMENT

The threat is never at rest, and never takes a holiday. Just because you have problems obtaining intelligence about any attacks or suspect persons or are unable to keep up with the ever changing tactics and techniques of the threat does not mean they are sitting there watching cartoons. Whenever you have no good intelligence or cannot obtain any actionable information at all a Personal Security Detail or company must use their combined experience, awareness of the operating environment, knowledge of enemy tactics and techniques, trust in their own tactics and skills and of course use good common sense to produce the best chance of mission success that is possible. You cannot sit in a secure location all the time, because your client has a job to do, and you must do yours.

Not every situation will be perfect or 100 percent safe but you are going to have to move the client and keep him safe to the best of your abilities. Sometimes you will have to cancel an operation or movement because of specific information that puts him in danger, but this should not happen frequently. You cannot paralyze your client and keep him from moving all the time. If your client can't do his job, then there is no reason for him to be there.

I had a situation in Mozul, Iraq where the subcontract security company had information that the threat had car bombs in the city. The subcontract security provider decided it was too much of threat to move. I asked his reasoning and his source. He got it off of a news broadcast, so I pointed out that there have been car bombs in the city for over two years and

we needed to get to the work site. He refused, saying the threat was too high. We conducted the necessary movement with no incident. When I got back I called the company's head boss in country and went over our contract, pointing out they could not cancel any run. It was up to the client to make that decision based upon the information provided to him. Using an international news service based in New York as your intelligence provider for local operations and then deciding you will not move the client is not what a security company is hired for! However, many get away with it because most client companies don't have their own security manager to oversee operations for them. They contract out all security operations and allow others that don't have the company's best interests in mind to make decisions, which is a major mistake in my point of view and experience. There is always going to be risk. If there wasn't why would they need a security detail? You have to learn to operate in a high risk environment; you do this by limiting the risks you take, through RISK MANAGEMENT.

So what is RISK MANAGEMENT? You could say it is a process of systematically identifying, assessing, and controlling risks that will arise when conducting protection operations, including personnel security, movement security and site or physical security, and then making the best possible decision with the information collected that will balance the risk and the cost of mitigating the risk with your client's operational or mission requirements. The final product of the risk management process should be the identification of those areas and assets that are vulnerable to the identified threat's means of attack.

The assessment of risk needs to be based upon several critical components of your proposed security program. These include a good threat assessment, asset and physical security assessment, and a vulnerability assessment. Then the security detail and the client need to decide which assets require the most protection, remembering people are more important than equipment, and where you need to plan for future expenditures to minimize the risk of attack or lessen the severity of the attack when or if it happens.

RISK ASSESSMENT AND REDUCTION

Now that we know a little about the potential threats, I want to talk about the more specific ways we can plan on mitigating that threat. The first way will be by conducting a review of assets so you know what you need to protect, a threat assessment, a physical security assessment and a vulnerability assessment. When you have that done you will combine that information and have a risk assessment. This will help you in your planning of client operations and in determining the level of security required for each location, and/or event, which in turn will help you to eliminate or reduce the threat's options of attack, based on the current and past actions of the potential threat. This is difficult to do because you have to remember the potential threat operates outside of your area of control so the potential threat is caused by factors outside of your bubble of influence.

The risk reduction process begins with your assessments of your client's security profile for personnel security and physical security of locations. I prefer not to look at previous assessments until I have conducted one of my own and can come up with my own conclusions because sometimes people's judgments become clouded or perceptions altered after reading someone else's security assessments. A really inept or lazy company will just make minor changes in an old security assessment and then present it as

their own. As the client I would not give access to any previous assessments until the security company hired has conducted their own and come up with their own conclusions.

I have used the terms threat assessment, asset review, physical security assessment, and vulnerability assessment, but basically you can call them anything you want and many people do. What you need to do is to identify the potential problem areas such as accessibility to the client at all locations and during movement, predictability of actions, and perceived vulnerabilities. Some examples of these types of things are:

- using the same routes for daily movements

- leaving and arriving at primary locations at the same times

- using the same vehicles

- always using the same airlines

- guards changing shifts at the same times

- physical security of a site

- deliveries always being made at the same times

- golfing or visiting restaurants regularly

- birthday parties held in the same place

One of the primary things you are looking for is anything that makes the client predictable or has become a routine even if it only takes place every 3 months.

After you do your assessments and/or surveys you will need to develop alternative courses of action for each perceived area of vulnerability. Then you need to war game these to see if they will meet the client's needs and are within his ability to change, because it is a big waste of time and effort to come up with a plan that the client cannot use. After the war gaming and discussion you will then need to evaluate the alternatives and do an assessment of these alternatives, from start to finish. If you propose new routes, you need to drive these routes and do a complete route assessment on each one. If you propose a change in any security procedures, you need to do a full evaluation before you make them permanent.

Sometimes you might want to bring in an outside company or another person to look at your changes and redo the assessment because many times they might catch something you missed. But remember you don't want to propose changes to current operating procedures without evaluating whether the new changes will meet the client's objectives and his operational needs. You need to make sure the changes are possible and will basically work.

After all that is done, then you have to select the best alternative plans and present these to the client with all the pros/cons and why the changes you are proposing are the best course of action for him (the client) not you (the security provider). Then it will be time to fully implement your new courses of action. After the new operating procedures are in place you will need to continually evaluate results of the new procedures. The security plan should be a living plan and you should always be looking for ways to refine it and finding ways to make it better.

The end result of your assessment or survey process should be able to tell you the following information:

1. **Most likely type of attack**. This will be based upon two things:

- Externally: how the potential threats operate, what resources they have, their past history of attacks.

- Internally: your vulnerabilities, areas of weakness, your available resources.

2. **Most likely place of attack**: Whether fixed site or during movement. Each potential attack site must meet the criteria for the threat to use as an attack site. THE VICTIM picks the place through his routines and security.

3. **Most likely times of attack**: This too is something that is picked by the VICTIM, through his routines and security. The time of the attack will take place when the victim is in a location that the threat can control, and it will take place when it is advantageous to the threat.

Your findings need to be clear, concise and to the point. If you spend too much time going into the detail of how you came to that conclusion, when you present your findings no one will be listening by the time you get to the part that matters.

There are examples of survey and assessment formats in the Appendix.

THREAT ASSESSMENTS

A threat assessment should be initially conducted to give you a starting point for designing your security plan for the client. It needs to be a living document as part of the constant process of gathering and assessing information about persons and organizations that for some reason have an interest in your client, motive against your client or a publicly discussed intention against your client, and also have the ability and resources for mounting an attack against your client. So the goal of any threat assessment will be to identify all potential threats to you and the client and do as much as possible to prevent an attack on your client by using all the information gained and avoiding the potential threats by not being at a time or place where the threat can conduct an attack.

During the threat assessment you will identify what potential threats there are against your client and his assets and what tactics could be used to attack those assets. This is what will help you design the correct security for the client and his assets.

So what is a threat to your client and to you? It is basically any situation and/or circumstance that can cause harm to your client or his assets. We have already discussed some of the potential threats and there are many more which could be problem areas to the client. So you will have to list the potential threats with the most serious at the top. Then you need to list threat tactics that have been used against their targets in the past, their capabilities of gathering intelligence and obtaining weapons and how this matches up to the client's areas of vulnerability.

There are six factors to consider when focusing on collecting and analyzing information from all your various sources on the potential threat. They are: who, what, where, when, why, and how. These can be broken down further into the follow:

- The **existence** of a potential threat is who the possible threat may be.

- The **capability** of the potential threat will answer what the possible threat may be.

- The **intentions** of possible threats will answer the question of why you are a target.

- The **history** of the different potential threats answers the question of what type of threat tactics will be used.

- **Targeting** answers the questions who is most likely to be attacked and why they are chosen.

- The **security environment** of your area of operations will be a separate modifying factor to the capabilities of the threat and how you need to modify your security operations to meet them.

VULNERABILITY ASSESSMENT

Next you will need to eliminate or reduce the vulnerability of the client in all areas. Vulnerability is based on our client's past and present actions and your actions when developing and running security. Vulnerability is not based upon the potential threat's actions but on your client; what is vulnerable, when, where and how to fix it. All of these and possibly more fall within the security bubble you need to create around your client, especially in areas where you have control over the environment and the threat has none.

Tactics that you decide to use will affect the threat in several ways including how they plan and what tactics they need to change to penetrate your security bubble. Your changes might even make the threat move to a softer less threatening or prepared target, but you need to remember one basic fact and that is most vulnerabilities are within our abilities to control. After all, vulnerability is based on the client's and our actions or inactions and in almost all instances, we can make our client less vulnerable and reduce the odds of them becoming a victim of the threat.

A major part of the process of reducing your client's vulnerability is target hardening. This is using proactive security measures, not limiting yourself to reactive ones. It basically means that you will not prevent an attack from the threat, but what you can and most probably will do is make them attack somewhere else, to a softer target, or make them use a tactic they are not comfortable with or attempt something beyond their comfort zone of operations, which will give you the chance to react and neutralize, or evade the attempted threat incident.

What is considered to be a vulnerability to your client? It is an area of weakness in any aspect of your client's security profile where the likelihood that a particular type of attack can take place and will have a harmful effect on your client or your client's business interests. There are several factors that need to be looked at when looking at vulnerabilities, including:

• **PREDICTABILITY**. How predictable is your client or your client's business? How hard is it for people outside the client's security bubble to know where he is going to be, when the client is going to be there, how long he will stay, when is he going to leave, how often he will visit a specific location or use the same route?

• **ACCESSIBILITY**. What is your client's accessibility? Your client or client's businesses can be harmed if the threat can get to them.

• **RESOURCES**. Do you have adequate resources? Having the appropriate resources is one of the basic problems of any company starting a new security program. Most companies do not feel the expense of a good security program is necessary until after an incident has taken place, lack of appropriate resources is something you will have to work with the client on and develop the necessary requirements to meet their needs for short and long term security of their assets and personnel.

• **PLANNING**. One more area we need to look at is poor planning, which happens a lot, especially in places where people (security companies) get complacent, and this in turn gives the client a false sense of security and he gets complacent. I have heard terms like "milk run", "the regular trip", "been there before nothing to worry about", "that area is safe", etc. All this leads to people getting complacent and lazy, not wanting to do their job, and usually their job as it is outlined specifically in the contract. You need to treat every trip the same. There are no safe areas, no milk runs, and there is always something to worry about; if there wasn't why would the client need security? You need to assume every trip has a threat, plan each one accordingly and with the same dedication. Do not get into the mode of just winging it, because you are too busy, lazy, feel it is safe or because you are trying to save your own company money.

RISK ASSESSMENT (COMBINING IT ALL)

CONDUCTING A RISK ASSESSMENT FOR SECURITY

Definition: Assessing and controlling the amount of risk in client's life, at work, at home and during movement. To find out what the risk is you must combine the potential threats and their capabilities with the potential vulnerable areas of the client which will give you an idea of the risk you will have to plan for. You must have at a minimum these two things to determine risk or you will not have enough information to set up a security program!!

After you have finished conducting the review of assets, a threat assessment, a vulnerability assessment, you are getting ready to come up with a security plan to manage the risk of your client.

Your final risk assessment will include a detailed review of the client and his activities, his movement, and his work, family and personal circumstances. This is done to determine what areas are weak in security and can be potential problems for the client as they will be areas that the potential threat can exploit. When looking at your client's areas of potential vulnerability and/or risk you need to consider lots of things, such as:

- **Fame or notoriety**. Is your client famous? If so, what for?

- What is your client's **public image** in the operation environment?

- Is your client extremely **wealthy and/ or powerful** and is this well known in the operating environment?

- Is your client conducting any **controversial activities** and/or programs such as working for a government or person that is not liked?

- What type of **environment** will the client be working, living and moving through?

- Will this in any way increase his **visibility** or become a reason for the threat to target him?

- What **associations** does your client have that can draw negative attention to him and/or his business? This can be important because you wouldn't want your client to become a target just because he associates with people or corporations that have already been targeted by the threat.

You are basically looking for any activity that will increase the likelihood your client could become a target. You client should be checked against all these factors, whether it is a person or a business, and he should help if possible in this portion of the assessment to give his views and thoughts on business associations and personal associations he has that may affect notoriety.

CHAPTER 8
THE ADVANCE

An advance allows the maximum protection and convenience for the client and a smooth transition and operating environment for the security detail. Overall you can say that an advance detail is responsible for getting the site to be visited ready to receive the client or clients before they get there. How smoothly the planned mission or operation is at a site or facility will depend on how well the advance is done. Usually you have some of your most experienced, trustworthy and can-do guys on the advance team.

The advance is responsible for site security at the operations site, including coordination with the local law enforcement, and military, having EOD do a security sweep, preparing radio equipment and telephones, securing the means to identify who is who, checking employees who work at the site, checking traffic into and out of the site, controlling the tactical high ground if there is any, and identifying evacuation routes from the site to a safe area or evacuation holding area for client pick-up. How much the advance can accomplish depends on resources and personnel assigned to it.

The advance responsibilities also include all those things internal to the operations site and/or facilities, including floor plans, elevators, seating of guests, handling the media, receiving lines, rooms lists for each person and organization's location, maintaining the keys with an accurate sign out log, emergency lighting and back up generators, fire fighting equipment, a holding room with communications for security violators and a safe room with communications for the client.

The advance should also look at the approaches to the site location including:

- All ingress/egress routes

- Escape routes

- Logging the travel times to the location from the nearest police facility, fire station, hospital

- Is the hospital open 24 hours? Does it have physicians or nurses and medics? Does it have a blood supply on hand? Does it have ambulance capabilities?

- Any scheduled events and other facilities that will be visited or might need to be used.

Internal to the site the advance needs to find out the information for the routes in and out of the building itself so the client and guests don't go wandering off. They will designate the motorcade arrival/departure area, what barricades are needed and their placement, public access restrictions, and designated parking areas and locations.

RESPONSIBILITIES OF THE ADVANCE

• Guides the motorcade into the site via radio and/or telephone communications, establishing visual contact with the detail leader as it gets closer to the debus site.

• Updates the shift leader on the current situation at the site itself and at the debus location.

• Meets the detail leader or bodyguard and invites the client to follow him/her to the client's destination.

• Sets the security perimeter, areas of observation, fields of fire, and prepositions guards.

• Employs EOD teams to conduct bomb searches of the areas to be visited.

• Prepares the debus/embus site for the convoy and locations for the convoy vehicles to stage awaiting departure or emergency evacuation of the client.

• Establishes a hard copy survey for possible close landing zones for helicopters, locations of all hospitals and clinics in the area, all defensible positions at the site, and the routes to and from all these locations. Verbally gives the location of the details on site safe havens for emergencies to the detail leader or bodyguard.

• Purchases before hand or obtains any entry passes, tickets, parking passes, and event badges so they can be distributed on arrival without any delay to the client or break in security.

• If the object of the mission is a meeting, the advance should establish contact with the person the client is going to meet and their security so it will move smoothly.

• Determines emergency evacuation routes from the site to a safe haven or to the convoy vehicles in case of an incident that cannot be contained.

• Finds the location of all rest rooms and other comfort facilities on site so he can take the client there as needed.

• If the mission includes visiting a restaurant, the advance will reserve seating for the client that meets his security requirements and he will obtain seating for the security detail, and arrange meals for drivers and outside personnel who will not be entering the site.

• The advance finds rooms that could be used for possible private meetings, discussions or a place the client could talk privately on the phone that are not part of the site's daily agenda. The advance will search these rooms for electronic surveillance and obtain the keys, if possible, to secure them.

• The advance must know the locations of all stairwells and emergency exits, how they operate, and where they lead to.

• The advance leader will pre-position his detail to provide a secure environment for the client. He will use low profile or undercover security team members as required.

• The advance must know the local security chief or leader, where his men are located, their ROE for incidents and proper uniforms and equipment of local security personnel. If possible obtain pictures of these personnel and distribute to the detail personnel who will be closest to those positions.

• The advance detail leader will give his recommendation to the detail leader and shift leader on the level of security required and the security posture of the detail.

• Provide a strip map of the site to the arriving detail to include any pertinent phone numbers or contact names.

A SAMPLE ADVANCE QUESTIONNAIRE:

• Anticipated arrival time of convoy?

• Exact location for the detail's vehicles to park?

• Will there be any VIP's or others there to greet the client as he arrives?

• Entry route to be used into the facility and to the meeting site?

• Is there a private elevator available? What is its holding capacity?

• Will a meeting room be used or will the meeting take place in a residence? Number of personnel the facility will hold?

• Was a bomb search conducted by an EOD detail and has a security sweep been done?

• Does the facility already have an emergency evacuation plan? Does it meet the security requirements for the client's detail?

• Where is the nearest fire fighting equipment located? Does it work?

• Type of ID badges, pins, or cards needed to gain access to all areas of facility?

PRE-ADVANCE WORK

One of the most critical parts of the advance is the pre advance work that is done. In this phase, you gain all the information you need for planning the movement and this can make the difference between a smooth uneventful trip, and one that ends up a little more exciting than it needed to be.

Information the pre-advance will need to have:

ITINERARY

Complete **itinerary** of the movement from the client including:

• All personnel who will be traveling (need this information for the security detail also).

• The dates and requested times of the movement (times will be adjusted by the detail leader for maximum protection but it is necessary to know the time the client is required to be at the location).

• Contact information for each individual going including home phone, cell phone and work phone, names of next of kin.

• If all the people who will be traveling with the security detail are not in the same location, what are their schedules of movement to the departure point of the convoy?

• Size of personnel traveling so you can have the correct size protective gear on hand if they do not have it already.

• Type of transportation the client has requested? Is it available? Does it meet the security requirements for the anticipated threat in the operations area?

The itinerary should be obtained and reviewed before any movement. The itinerary for the movement and the event agenda can usually be obtained from the client. But sometimes project officers or event coordinators who are hosting the event will make their own itinerary, so the advance will need to obtain a copy of this also; it will be important to know if there are two itineraries or event agendas and to find out how they compare with each other.

If there are differences, the client will need to be brought up to speed to determine if the new itinerary or agenda meets with his needs and requirements. This is necessary because sometimes project officers, site managers, and hosts will go beyond what was requested for the event by the client. When you compare the different itineraries, you can correct any problems and misunderstandings beforehand thus preventing an embarrassing situation for the client and the host.

Each site and event itinerary and agenda should have identified points of contact (POC), with room numbers, building locations, phone numbers and email addresses. Having this information will allow you to know the POC in advance, so you can schedule any appointments before you leave to begin the advance work. Remember, you don't want to just show up unannounced or unexpected.

When finding out about the mission, the itinerary should show you what kind of event the client or clients will be attending. This in turn will determine what clothing will be worn by security personnel, especially the advance so they can blend in when setting up the static security, and what will be worn by the shift leader/bodyguard so he will blend in while with the client. It is very important that during the advance, the personnel at the site are asked what clothing they will be wearing for the event, and what clothing the site staff will be wearing. There are few things more embarrassing than when a detail and a client show up at an event or site dressed differently than the hosts and other people attending the event. Security personnel that will be inside the facility

during the event should be dressed in the same way (casual, formal) as the client. Those outside, depending on the threat environment, should blend in with the site personnel.

The itinerary or agenda should tell you if any special equipment or clothing will be needed for you, the bodyguard and the client or clients. If your daily agenda calls for visiting a power plant or construction site, having safety helmets available for everyone will be necessary. If the client or clients will be visiting a live fire range, safety equipment will need to be obtained. If going on a boat or yacht, a life preserver, sun block, hat, and UV sunglasses might need to be obtained. What if the area or site your client will be visiting is a high threat area? Special weapons and ballistic vests might need to be obtained, if you are working in a permissive environment. This is one of the reasons why it is important to know the sizes of your clients and keep that information on record. It will shorten the time required to track down the information and obtain the necessary equipment to make sure your client or clients can attend the necessary events.

The entire itinerary should be reviewed as early as possible for two main reasons. The first is to make a determination if there is enough time to conduct an advance. There will be times when itineraries and/or agendas are packed with multiple events, trips, dinners and/or meetings. When this is the case the advance needs to head out as soon as possible and use the additional time to complete the advances to the different sites, locations and/or venues for that mission. The second reason the itinerary should be reviewed early is to determine if there will be sufficient time to accomplish the itinerary timeline as it pertains to the travel times.

TIME CONSTRAINTS

How many days should be necessary for advance work?

This is a question that many experienced people could answer many different ways. One rule is you need a minimum of 3 (three) working days for each operational day of the mission if the client will be traveling to multiple sites. If you are going to one facility for the entire 2 day mission then 3 days should be more than enough, depending on the operating environment and threat. So at a minimum 3 days should be calculated for an advance team to do its job the first time it works at a site. If it is a site that is visited often, like a hotel at the airport or a facility used by the client company, then after the initial advance, unless there are massive changes at that location, arriving 4 hours prior to the client should be enough time to set up the area and get security in place. Sometimes, arriving 30 minutes prior might be enough. It all depends on the location, knowledge of that location, threat, who controls the location when the client is not there, etc.

Remember, however, the minimum time should be increased or decreased based on the amount of sites listed on a day's itinerary, the time and distances to the sites, the complexity of the site, the daily agenda, and the threat. The scope of the advance will also depend on resources available and personnel available for the mission.

The calculation for the amount of time required for a mission can be calculated from the following formula:

TIME REQUIREMENT STANDARDS

Standard Locations: Allow 2 hours per each separate location listed on itinerary. Multiple offices in the same building can be considered separate locations and each requires a separate site survey done for that office.

Special Event Locations: Allow 4 hours per each separate special event location listed on the itinerary or agenda. Special event locations can include, but are not limited to, sports stadiums, concert halls, parade routes, amusement parks, malls, parks/public memorials, any location containing numerous public entrances/exits, and any event with large crowds.

When the advance is conducting route recons and surveys you should schedule about 6 times the amount of time required to travel one-way from the origination location to the destination location during the time of day which the detail with the client will be traveling. This is so the advance will have enough time to conduct surveys of the primary and alternate routes in both directions, and allow time to search & coordinate safe havens, check danger areas and choke points.

So how much is enough time for the advance to set up and prepare an area for the security detail's arrival with the client? There is no way to answer this since there are so many variables involved. It will depend on many factors, including whether a site advance was done here before, known contacts, familiarity with the host and his security company, and good liaison with the site but as a general rule the advance should arrive a minimum of one hour prior to the client's arrival. However, the minimum time should be increased for more complex sites or more difficult itineraries or agendas, or if there have been any major changes to this site since the last visit by the security detail.

Also if the person who is doing the advance this time is not the person who conducted the survey, you need to plan additional time to allow the advance to conduct a proper advance and be familiar with the entry, exits, safe haven, holding room, and the site before the arrival of the client.

LOCATION

Every personal security company should maintain after action reports, mission planning and other historical files concerning each place visited by them, regardless of who the client was during the visit. This should give you POCs, contact information, routes, site and facility information. You should keep separate files on current clients and sites they have visited. This way, when working in an area for a prolonged period your company will build up a good database of information that can be used for future planning regardless of who the client is. Why reinvent the wheel? This way you can use previous surveys to your advantage. By reviewing these files, you will have a general idea of the routes, location and layout before you even hit the ground. You may even have the same POCs. Also by reviewing AAR's of previous missions to this site you can determine what, if any, problems occurred and possibly correct those problems for your mission.

The security company should have area files and country files for any location you may be traveling to so the detail leader needs to get these files to review. In this file will be some basic history about the country and its customs and courtesies. I know most people are saying Arabic is Arabic or that working in a country things are the same all over, but this is FALSE. Think about the differences in working in Baghdad and Mozul, or any area where tribal customs or local ethnic customs will differ from the normal environment you are used to working in. It is very important to know what the customs and courtesies are for the area you will be visiting, what language is spoken, etc. It would not be a good thing if you or your client found yourself in jail for an offense that is not against the law in the United States, or for doing something that is permissible in another location like firing warning shots at vehicles that get too close in Baghdad. I encountered this problem with a detail that traveled to Erbil. The tail gunner fired at a vehicle and the detail was stopped and surrounded by Kurdish law enforcement and military.

In some places you might need a special permit to carry a weapon or written permission. This can be important. For example, it is against the law to carry or possess a gun in England. Why do you need to know this? What if your flight or mission forces you to fly there to transfer airplanes? The security in British airports is set up so all luggage is screened again regardless of layover. When a plane lands in England, having weapons or other types of equipment in your luggage will put you in jail. Not good for you, very bad for a client.

The detail leader, but most especially the advance leader, should learn the local and host country holiday schedule. This could be something that could affect your client's mission, and force you to make changes to the security program for the time of mission. The holiday event schedule should also be researched to find out about parades, plays, or sporting events that could affect the mission. Learning about these will determine if additional time may need to be planned for advance work due to a holiday. An excellent example is Ramadan, the Muslim Holy Holiday in the Middle East. During this time your advance time will be shortened because of the prayer schedule followed by all followers of that faith. Also many restaurants will be closed and chances to dine are limited since there is a fast involved. Some countries celebrate their weekends on different days than Western Cultures, for example the weekends in Saudi Arabia are on Friday and Saturday. Don't forget commercial events too. Carnival time in Rio would be difficult time to plan a mission for a client.

Before the advance departs on its mission, you need to learn if there are any weather restrictions for the mission location or along the planned routes of movement. Some places, depending on terrain, may have extreme fog during the morning hours, monsoon seasons, or snow in winter (especially if the mission is to an elevated location), high winds or sand storms. You should also learn what the normal climate is for that area prior to departure and what the forecast is for the time of the mission. After all, you could find yourself going from a hot climate to a cold winter one, just by moving to different elevations in the same country. If you are traveling between different countries this could be even more extreme, especially if you have to do a back-to-back trip. In that case you will have to pack accordingly for the different climates. It is also advisable to have an operations fund in case the advance needs to purchase clothing for unusual weather conditions for themselves, the security detail and the client.

You don't have to move between countries for weather changes. We moved from Baghdad to Talill in southern Iraq to do a site survey and physical security assessment, and half the security detail froze at night because the temperature dropped way down compared to Baghdad and because the tents had no heaters.

Keeping your vaccinations updated is very important because some countries may require you to have certain shots and show your vaccination records before entering. An example is some African nations that require you to show your vaccination records and have proof of a yellow fever shot when going through customs. For some areas in a country you may have to take certain medications such as malaria pills while you are working in that area. Before you leave on a mission, check with your in-house medical staff, at the local US embassy or with the military to determine if any vaccinations or medication are suggested for working in that area. Regardless of how often you work overseas or where you work, you should always keep up with all your vaccinations and always travel with your current records.

THREAT LEVELS

There are 3 levels of threat that are normally used: High, Medium and Low. The advance should always obtain the local threat level before departing. You can get this from the government or military, other security companies working in the area, or in-house since most private security companies have an intelligence section working in their operations center. You can also contact the local area law enforcement for current threat levels and if going overseas, you can contact the RSO's office at the state department. Knowing what others think the threat level is for the mission operating area will help you determine if additional security needs to be added and additional precautions need to be taken.

But once on the ground, one of the most important things the advance needs to do is determine the ground reality:

- **What is the threat to the client?** They will need to figure out what type of threat there is, is it credible, is it directed towards a person or a general group of people based on their nationality.

- **What is the source of the threat information being used to determine the threat?** The advance usually gets threat information from their Intel Officer, and a variety of agencies: MI, Provost Marshal, Local Law Enforcement, State Police, FBI, Secret Service, CID, OSI, NCIS, RSO, etc.

- **Is there a specific threat against your client or client company?** This could mean that your client will have a threat level that is different from others operating in the area. If this is the case then you will need to brief him, the client company and his staff immediately. It may change the mission.

This collection of information will help you determine the type of security you will need to protect the client for that mission.

The threat level helps to determine the number of security people needed to operate in that mission environment. If you are traveling through or to a high threat area you may want to go with low profile vehicles with a security advance patrol and a counter attack team. In a hostile fire area you may decide to go with a high profile movement and be very aggressive. Make your decision based upon the need to keep the client safe and secure, not on what looks cool or the way it's always done by your company.

SITE SPECIFIC REQUIREMENTS

Some countries require you to have a valid passport and a country visa before you arrive there. Although most allow you to obtain a visa at your point of entry, others require you to have one before you arrive, and these can take several weeks to obtain. Many countries do not allow the bringing in or carrying of weapons while in their country. You need to coordinate with the appropriate authorities and provide all necessary information required to receive this permission. This doesn't only go for countries; there are some facilities, sites, and areas where they have their own rules and regulations. A weapons permit issued by the ministry of the interior for Iraq is no good in the north, because Kurdistan issues its own. Some military controlled areas do not allow any civilians to carry weapons within its boundaries. It doesn't matter who you are or what you have signed. That area commander has the last word regardless of nation of origin.

EXAMPLE: In southern Iraq, once we stopped to eat and rest at an Italian base where they did not stop us going in but attempted to confiscate our weapons when we were leaving. Luckily an American patrol was behind us and intervened, saying they always have that problem with that base, so know the rules, follow them and have alternate plans to meet your mission requirements as necessary. Remember FLEXIBILITY is always the key!!

Knowing what will be required to gain access to facilities and venues and what method of payment is accepted can be very important. I have seen a mission almost stopped because the hotel they had arranged for everyone to stay at and the one that had been cleared did not take the credit card they had to pay with. Luckily the client paid the entire bill on his corporate card, with the security company reimbursing him after the mission. This was very embarrassing for the security company and made them look like they did not have a clue what was going on. Some areas or countries do not take American currency and will only take payment in cash.

It might be advisable to hire an expeditor whether you are traveling to a different country or to another city. Expeditors are people who can smooth the way for you. They have all the necessary contacts and connections to help you bypass red tape or other problems that may arise when trying to get a mission ready. They can be expensive, and you might have to ask around before leaving on the advance to find a good one. Usually you can get a name from another company that does business there or the client's headquarters may have a local office or POC to the location you are traveling to, or the embassy usually knows who the expeditors are in their area. An expeditor is not cheap, but usually worth the money, especially on a first visit when you know no one and have no local contacts.

OTHER CONSIDERATIONS

The security detail leader will need to determine what personnel will be going on a mission for the client. Remember your client will probably be a company not a person, so besides the primary person you will be moving and setting up for other people including a spouse, family members, subcontractors required for the meeting or the client's staff. The number of personnel involved determines:

- How many vehicles will be needed for the movement.

- The seating plan for the movement.

- How many rooms to get for overnight stays

- If the advance needs to locate both male and female toilet facilities.

- How much extra equipment will be needed, such as body armor if people accompanying the client do not have their own.

- The seating arrangements at events.

- If any additional security may be needed such as a PSO for the wife or another VIP accompanying the main client.

- If female security members are needed.

The advance needs to find out if there are any other personnel or VIPs who will be staying at the same hotel or attending the same functions as your client. The host facility security manager should have that information, and you will need to work with other security companies as much as possible to make sure everything runs smoothly. You can determine who has the most experience at that facility or in that area of operations and allow him to take a senior role in coordinating all the security details and advances. This can be necessary so as not to cause an embarrassing situation with different security companies arguing or fighting each other for position, or trying to bully each other into doing it their way. Mutual cooperation is a

must and will make the whole event more secure.

This coordination will also ensure that you are not blocking each other's radio frequencies, and can identify other security company team members by a lapel pin or an ID badge so you know who is supposed to have weapons and who isn't, and who is supposed to be in certain restricted access areas and who isn't. NOTE: I always carried about 50 lapel pins of one certain type; whenever we worked with other details or security people everyone got a pin so they could be readily identifiable to each other.

Depending on the itinerary or agenda of the event being attended you need to identify any support elements that you may need prior to the mission, such as dog support for EOD sweeps, technical support for Technical Surveillance Countermeasure Sweeps (TSCM), Police support and any special equipment support. If you are going to need EOD sweeps done then the dog support should be coordinated for all missions and sites if possible. Also TSCM sweeps are necessary to ensure the privacy of the client and any business discussions that may take place; they can be done on the hotel room or special meeting rooms where the client will be located. Outside support from local law enforcement may also need to be coordinated for additional security and/or traffic control at the site.

You need to identify any special skills that may be needed for the mission, for the security portion AND the client, because you may have to coordinate to obtain an interpreter for the meetings or a transcriber. Or if there are going to be recreational activities conducted, you may need a ski or scuba instructor or tour guide for the local area.

KEYS TO SUCCESSFUL ADVANCE WORK

• **Do not confuse your position with the client's rank or status**: Don't use your client's position or rank to demand things. Remember you are a representative of the client, but you are not the client.

• **Be prepared**: Know what needs to be done and what questions you want to ask beforehand.

• **Make Contact:** Make contact with personnel at the site to set up appointments and introductions before you go to a site. Do not just show up unannounced, or without prior warning to the host or the host's security detail. It's annoying and unprofessional.

• **Punctuality**: Be punctual. Try to get to the appointment at least ten minutes prior. If you are going to be late, call them.

• **Have the Proper Attire**: You should always have the proper attire for the environment you will be working in and where your client will be working. It is not always going to be khakis, desert BDUs and full body armor and combat kit. Sometimes a suit or slacks and a sports coat with tie will be needed, or maybe just slacks and a collared shirt. You must take into account who the client is meeting, why they are meeting, the area where he is visiting and the way people dress there. (Some restaurants or other venues have dress codes.)

• **Establish good rapport**: You will need to establish a good rapport with your counterpart, locals and others you will be working with on the advance. You should always establish some rapport first before going directly into your questions and business.

• **Be flexible in your expectations**, not demanding. Be prepared that things will not always go your way and you will not get everything you want or think you need when dealing with people in different locations. Remember they have their concerns and have to do what's best for them and/or their client. Be as flexible as possible, don't make demands unless there is a direct threat to a client's security, and if something cannot be worked out, inform the client that the operation may have to be canceled or postponed.

• **Leave a means of contact**: Always leave a contact phone number with your point of contact at the destination site. Leave alternates too, in case your phone does not work. This is done in case there are any last minute changes at the event, before your client arrives.

• **Communicate with the security detail**: Always keep the lines of communication open with the security operations center, the shift leader and the detail leader. As the advance you are the eyes and ears for the mission and the detail as a whole prior to their arrival at the site with the client.

The morning of the mission you will need to do several things:

• Check on any **last minute changes** at the site or for the agenda, so you can coordinate with the staff at the location and the person in charge of the agenda for any last minute changes prior to the arrival of the detail with the client.

• **Security operations center**: The local security operations center should be up and running with all communications checks completed about two hours prior to the arrival of the detail with the client.

• On the day of the mission, if meeting the security detail at an **airport or train terminal**, the advance needs to arrive a minimum of one hour prior to the arrival of the client. The advance should also know the airport and how customs or security works and if possible meet the client/detail at the gate and walk them through.

• The advance leader should contact the **security operations room** each time the client moves to a new location at the facility or site giving them location information and any other details they may need.

• **Coordination** between the shift leader and the advance leader should be continuous until the arrival of the client at any site or venue visited. The advance leader will inform the shift leader prior to arrival who will be at the debus site to meet the client so he will not be surprised or caught off guard. This includes informing the shift leader of any crowds, protesters or media present. If the advance leader is not going to be there for the arrival, then he will inform the shift leader of who in the advance will be there to take his place.

• The **advance leader will be outside** in a visible location so he can provide direction to the convoy as it is arriving.

• The advance leader will take charge of **moving the client around the site** since he has the most knowledge of the location and should know what is going on. The detail leader/bodyguard will remain with the client and perform his duties as a close protection officer.

• The advance leader will make **introductions** as necessary to the client, since every person wanting to be introduced to or who are on the schedule to meet the client should have been contacted and had a meeting with the advance prior to the client's arrival.

• The advance detail will assume previously arranged **security posts** to assist the PSD team in setting up and maintaining a 360 degree security bubble for the client.

• The advance team **remains on site** until they are released by the shift leader who is the overall tactical commander of the client's security detail.

<u>*NOTES*</u>   <u>*NOTES*</u>

CHAPTER 9
SURVEILLANCE DETECTION PROGRAM

The goal of a surveillance detection program is to detect any surveillance taking place against the client, the client's property and the security details protecting them. A surveillance detection program is nothing more then training people to be surveillance aware. Why is this important? Every successful attack against a target, especially one with security in place, whether it is an assassination, kidnap, etc is the result of careful planning and intelligence gathering on the part of the attackers. They get their intelligence through several means but the most reliable and best intelligence they get is from conducting surveillance on their planned target. There are many different definitions of surveillance but it is basically "the systematic observation of a person, vehicle or location by technical or non-technical means to gather information on those being observed that cannot be obtained by any other means." By having everyone in the security detail surveillance aware you can reduce the chance of a planned attack against your client and his property.

So who uses surveillance? Well, everyone: the threat, criminals, foreign intelligence agencies, host nation military or law enforcement, the media, competing companies working against your client, etc.

What are their goals?

- The **threat** will want to attack in some way: kidnap, assassinate, steal, blackmail or commit some other illegal act.

- **Foreign intelligence agencies** watch everyone and your client may be doing something that is of interest to them.

- **Host nation military or LEA** could be keeping an eye on you and your client just to keep track of you and make sure you are not violating any laws.

- The **media** watches for the story of course, either about you or the client and the more outlandish or worse the information, the better the story.

- **Competing companies** engage in industrial espionage. Finding out how your client operates, how they bid, etc. is big money to another company.

Why will others carry out surveillance against the security company as well as the client? Because they need information on you, especially the potential threat. A threat needs to know what you're manning, what type of equipment and vehicles you use, your methods of operations, your capabilities, and your level of alertness both during the duty day and off hours. Surveillance against you and the client will not only take place at work or during work hours but will be continuous. It is much easier to conduct surveillance, especially close surveillance, if the target of that surveillance is relaxed and less aware of their surroundings. Depending on why the surveillance is being conducted, off time is more likely to be interesting to the media or the criminal for the purpose of blackmail or headlines.

So a significant portion of your overall security effort should be in countering any surveillance performed by the threat, hostile or otherwise. You can do this by knowing the different types of surveillance and how they are used and by performing surveillance detection during all missions and at all permanent facilities and temporary static sites. By randomly putting out counter surveillance and fixed point counter surveillance to help in detecting the threat. By always remaining alert for the signs and indications of surveillance you can increase your chances of spotting even the best trained surveillants. Remember the environment you are working in and the threat, these two factors will have great influence on the surveillance conducted against you and your client.

You can add a set of eyes to your surveillance detection capabilities by teaching your client these things also. You must be unpredictable in everything you do, even day-to-day things that cannot be avoided. Vary routes, times, vehicles, departure points, arrival points, etc. Since surveillance is usually the FIRST VISIBLE sign of a threat, the advantages to being able to detect surveillance or to confirm you are not under surveillance are many.

Detecting hostile surveillance provides you an early warning that an attack is possible, and by how professional the surveillance is you might even be able to tell how far along they are in the planning. This warning will let you know to raise your awareness levels and check your operations security. With that information you can make an informed decision (after briefing the client) on whether to continue the operation or not. If surveillance is detected, this will allow you to increase security measures or change the security profile as you deem necessary to raise your security level to sufficiently counter the threat whether it is against a person, place or thing.

TYPES OF SURVEILLANCE

Surveillance can be conducted many different ways, depending on a number of different criteria such as:

- Who is carrying out the surveillance?

- How well trained are the surveillants?

- Who trained them?

- What resources are available to them?

- What type of environment are they conducting the surveillance in?

- What country or area are they conducting the surveillance in?

- Are they concerned about being detected or compromised?

- What is the reason for the surveillance?

COVERT surveillance is usually the most difficult type to conduct so it is carried out by trained operators with experience. It can be used throughout the full range of surveillance methods and is extremely difficult to detect without the use of counter-surveillance and surveillance detection drills.

DISCREET surveillance can be carried out by people with or without training, and is used a lot by security forces, military and law enforcement. It can be used in the full range of surveillance methods. People conducting discreet surveillance are not very worried about being detected or compromised because they usually have a legitimate reason to be conducting surveillance and are part of an official law enforcement or military.

BLATANT surveillance can be carried out by trained or untrained operators and by military or law enforcement in plain clothes or while in uniform. When this type of surveillance is used the surveillants are not concerned about being detected or compromised. In fact they want to be seen, to antagonize the target of the surveillance and get him to commit an act for which he can be arrested or detained. The media likes to use this type of surveillance to force people to do things to make the news.

COMMUNITY surveillance can be conducted by any member of the community who has sympathies towards the person or organization that is trying to gather information against someone. The people performing this type of surveillance are usually untrained and may receive monetary payments or other types of incentives for passing along information. They could be anyone in the community you will be working in including members of the client's locally employed staff, vendors or restaurant staff. Sometimes they will be passing information because they are under threat of harm or other duress from the element seeking to gain intelligence about you and your client.

METHODS OF SURVEILLANCE

Surveillance can be conducted in many different ways, and each method can be used either alone on in combination with other methods to enhance the gathering of information. The training and resources of the person conducting the surveillance will have a big influence on the methods they can employ. The best way to determine the method that will be used against you and your client is to study the history of the various threats in that country to see what they have used in the past when carrying out attacks against targets.

STATIC SURVEILLANCE

STATIC surveillance means staying in one place when conducting surveillance. When it is done on a temporary basis, it is because the target has made a temporary halt, such as during a vehicle movement, refueling, shopping, rest break or at a mission site that is being visited. A surveillance team will attempt to use what cover is available in the surrounding area when this happens. In an urban environment they will window shop, use a public phone, or sit at an outside café. In a rural environment they can use hedgerows, bushes, trees, farm equipment, groves of trees, or roads that take them away from the stop but shield them from observation by the target.

During temporary stops surveillance teams will use this time to also plan for their next move. They need to plan and prepare to cover the target when they move from their stop, regardless of the direction the target takes. This can be done by pre-positioning their assets, either foot or vehicle, to cover all routes from the stop. If they have a position where they can observe the target from a distance, they can direct surveillance teams when the target starts moving again.

LONG TERM STATIC SURVEILLANCE

Long term STATIC sites are used when there is a stationary site to observe, such as a residence, work compound, or other primary area that the client needs to visit. These types of surveillance positions usually have 24 hour visual coverage of the target, blend in to the operational environment and have easy access to the surveillance location. This type of surveillance is extremely difficult to detect.

What do you need to look for when trying to detect static surveillance?

• First you are looking for any **unnatural behavior in individuals** who are in the surrounding area. Do they look uncomfortable or out of place for the environment? Why? What are they doing or NOT doing?

• You also are looking for the presence of anything **abnormal for that environment**. You should have a good grasp of what is normal for the environment you are working in so look around. Is there anything that is out of place? Anything not quite right for that area? Then ask the question *why*.

• In **rural environments** you need to study the area and environment to see if something catches your eye for some reason. Look for movement and noise in places where there shouldn't be any normally. Look at the shape of objects in the woods or hedgerows. Look for shine, correct shadows based upon shape of objects in the area, silhouettes and outlines and the different road or ground surfaces for anything that does not look normal.

• Observe for **correlating movements and activities** based upon your movements. Do other vehicles or people seem to be moving when you move? Do they change positions based upon your movement?

FOOT SURVEILLANCE

FOOT surveillance is usually conducted when the target of surveillance is moving on foot. This type of surveillance can be performed with anywhere from 1 to 6 surveillants. How many is used depends on the training of the people conducting the surveillance and the pedestrian traffic density in the area where the surveillance is taking place. For example, a high density urban environment such as a major city compared to a low density rural environment.

What to look for when trying to detect a threat who is conducting foot surveillance?

• They will need to communicate, which is much harder to detect now with the use of cell phones being so prevalent but you should look for people whose conversations correlate to movements of the target.

• They need to be in a location that allows them to SEE the target so look for people who always seem to be in positions of observation as it relates to you or the client.

Also, personnel conducting surveillance will try to stay out of the target's or security's peripheral vision to avoid unconsciously alerting the target or the security detail. Once again look for any unnatural behavior of individuals in the area, people who are not moving with the normal flow of pedestrian traffic around them. Look for people who seem to be uncomfortable in the area or look out of place for that environment. Also, pay attention to people who have the deer in the headlights look when you look at them or who look away when you look at them. All surveillants have an unconscious fear of being detected and will react different ways when a member of the detail looks at them. Be aware of people who seem to be loitering in an area for no reason and people who have no purpose for the action they are performing. Always look for the presence of anything abnormal to the environment you are working in. Once you find something try to figure out why it is out of place.

MOBILE SURVEILLANCE

MOBILE surveillance is conducted when the target is moving by vehicle. It can be conducted with one vehicle, which is very difficult, or up to 6 which is what is recommended. The team for a vehicle conducting surveillance usually consists of 2 people: the driver and one other. This allows the driver to concentrate on driving and the other person to take notes, photographs, and video and to get out and follow on foot if necessary. The vehicle used should fit into the operating environment, and probably be local, very innocuous, and drab colored so it doesn't stand out enough to catch the target's or the security detail's attention. The surveillance team will usually have all routes from a departing location covered by a vehicle so that whatever route you choose a vehicle will pick up the target and direct the other vehicles on the route and direction the target is taking (or some other variation of this). A trained and aware surveillance team will usually hand off the target to another vehicle after 3 major decision points are passed. A decision point is a location where a decision had to be made, get off a road or make a turn. The vehicle immediately behind the target will turn off after passing three of these and fall farther back and rejoin the surveillance detail, while the next car falls in and continues.

What you should be looking for is activities that correlate to the movement of your departure and your convoy. This includes:

• Vehicles that had been sitting for a long time then pulling out when you do, or people leaving areas and getting into cars as you pass and pulling in behind you.

• People who follow you through more than 3 decision points.

• Seeing the same vehicle and people in it more than once in areas separated by time and distance.

• People photographing or video taping the convoy from behind.

• People looking away or covering their face when the convoy passes them.

• Unusual or abnormal behavior by a driver or his passenger.

• Erratic driving by any vehicle as it tries to stay in position behind the convoy.

• Things such as peeking often. Peeking is when a vehicle is between you and the surveillance, the surveillance vehicle will move slightly to the side to "peek" around the car or cars separating you. This is usually done by inexperienced or untrained surveillants who are worried about losing you when you are out of sight.

• You should also look for people who are hanging back from the convoy/motorcade if they have the ability to pass, or a vehicle that "hangs back" when there are no vehicles between you and it and you are traveling below the speed limit. Once again this is usually done by inexperienced or untrained surveillants or surveillants that feel uncomfortable in the traveling environment and is usually because the surveillants are feeling overly exposed to the target vehicle.

A well trained or experienced surveillance team will adopt a natural riding position on the road in relation to the target vehicle.

TECHNICAL SURVEILLANCE

TECHNICAL surveillance is exactly what it implies. There are a wide range of technical devices available today including those for video, photographs, communications scanning, vehicle tracking, people tracking and audio surveillance. The purchase and use of such devices is not limited to government or law enforcement agencies. They are commercially available to just about anyone.

What to look for when trying to detect technical surveillance depends on the environment. Some technical surveillance devices need to be physically placed near the target of the surveillance or physically on the vehicle, so you should look for any signs of ground disturbance around the fence, walls, or hedges. When searching your vehicle, look for areas of disturbance around it, such as the ground underneath being disturbed or hand prints on parts of the vehicle that normally don't have them. Also some wireless surveillance devices may interfere with communications equipment, television or commercial radio signals so if there is sudden and prolonged interference in these signals this could be an indication of their use in the area. If you are using unsecured radios in your security communications you should use codes or find secure radios as soon as possible since off the shelf scanners can grab most unsecured signals.

AERIAL SURVEILLANCE

AERIAL surveillance is probably not one you will have to worry about in most places, but some areas or countries could be sympathetic to the threat. And some military or law enforcement could work directly for the threat so there is always a chance this could happen. Or a competing company could be using aerial surveillance to track client's movements and/or operating techniques.

The effectiveness of threat or hostile surveillance against you and the client will depend on:

• Their ability to **blend into the operational environment** well enough to perform the mission.

• Their **local area knowledge** of the operating environment, people, customs, language, and the type of training they have received if any.

• The **manpower** available to them for the operation.

• The **resources** available such as money, equipment, vehicles.

• The type of **support** that the threat has in the area of operation either active or passive and who the support is coming from will also be a big factor in how they operate in the area.

Successful surveillance, especially the covert type, requires that the target of the surveillance never suspects or sees the surveillance taking place. This is why we conduct surveillance detection operations, to help detect surveillance on us or the target regardless of who is performing it.

ANTI-SURVEILLANCE TECHNIQUES

Surveillance detection or anti-surveillance consists of proactive actions that are taken by a surveillance target to force anyone conducting surveillance on them to expose themselves or react in such a way that the target can determine if he is under surveillance. In extreme circumstances there are actions you can take to evade surveillance although this is not recommended in most cases. There is also a difference in detecting mobile surveillance and fixed site or static surveillance. First we will talk about mobile detection.

Actively detecting surveillance during a movement can be done several ways. The first series of ways are things you physically do and then watch for a reaction from potential surveillance personnel. But before you do anything you have to decide what is the desired outcome of your surveillance detection operation. Do you want to detect surveillance without those conducting surveillance knowing? Do you want to detect it in an overt way so they know you know and to scare off any surveillance? Do you want to evade any surveillance without it looking like you are evading it or do want to evade it so they know what you are trying to evade them? Once that is decided you next have to inform the security detail what is being done so they can raise their alertness level and look for correlating activities or reactions in the surrounding area of observation.

Before you leave for ANY movement the exit/entrance you are going out of, the route you are going to use and the area around your departure point need to be observed for 15 minutes, preferably by members of the detail. This way members will be able to recognize any activities by people or vehicles that are correlating their movements to those of the detail as they approach or pass.

The first surveillance detection/anti-surveillance technique is a **FALSE START**. This is when the convoy leaves the departure area goes to the end of the block turns around and comes back. Anyone who is conducting surveillance would have started the transition to the mobile mode: started a car, pulled out ahead, left a restaurant and got into a car. Now if they don't continue on with the action they will stand out as doing something abnormal to an observer.

The use of **DUMMY VEHICLES** is just like it sounds, sending out a convoy without the client in it. This convoy then drives a different route. While this dummy convoy is departing you will have people observing the departure area and the first part of the route for correlating movement or other activities started by the passage of the dummy vehicles.

The use of **UNSCHEDULED STOPS** along your route of movement will cause any surviellants to react to an unforeseen action. They will have to stop also and find a reason for stopping, which can cause them to get sloppy and make mistakes.

While traveling, **VARYING THE SPEED** of movement is done either on foot or in a vehicle. Go slower compared to the traffic around you and most people and vehicles will pass you. If one does not and maintains distance with the reduced speed there is a reason the person or vehicle is staying back there. The same is done if you speed up. If the suspected surveillance maintains the same distance from you then there is a reason. If you combine the two, speeding up and slowing down, and a person or vehicle always maintains the same distance from you then you probably have detected surveillance.

Use **ON AND OFF RAMPS** along major avenues of travel. You can take an off ramp during your movement then drive over the road and use the highway's on ramp to get right back on the highway in the same direction of travel. Anyone doing surveillance or following you would have taken the same off ramp. Now if they continue to follow you back onto the highway then the surveillance will stand out as an abnormal act.

Use **CONSECUTIVE TURNS**. In an urban environment you can turn 3 left turns in a row or 3 right turns in a row. Anyone following you conducting surveillance will have to make the same turns with you. Since you know why you are doing this, another vehicle doing it will look abnormal. Also it will be difficult for another surveillance vehicle to take over since you will be able to see it attempt to join in. This attempt will be made when the detected vehicle turns out off the surveillance line of the target vehicle so any vehicle pulling in to replace it should be considered possible surveillance.

DRIVE THROUGHS are when you pass through housing areas that have little or no traffic throughout the day, usually higher class neighborhoods, so any vehicles that look out of place for that area should look to suspicious to you. Driving through different types of areas that the threat might not have planned for when developing their cover will also make the personnel uncomfortable in what they are doing and heighten the chance of them making a mistake. Don't make this a habit or routine or they will get used to it and be prepared for it.

TRAFFIC CIRCLES (ROUNDABOUTS) are normally interesting places anyway since most people don't know how to use them, but when you add the element of someone trying to follow you without being noticed, it gets really interesting. A convoy can just drive completely around the traffic circle once before going back around and exiting at their exit. Since most people exit when they come upon their turn the first time, anyone who follows the detail around the traffic circle then takes the same exit will really stand out, because they didn't take that exit the first time they passed. A surveillance team will have to follow the target around the entire circle unless they already know what exit the target will take. That's why its important to have more than one route and to alternate use of these routes to avoid predictability.

TRAFFIC LIGHTS can be used just like in the movies. Last minute changes of lanes to move into a turning lane will require the surveillance to attempt the same thing before they get to the light. Also, you can slow down or delay going through a traffic light until right before it turns yellow then go thru before it turns red. Look for anyone who tries to run the light after you.

CUL-DE-SACS are basically dead end streets that don't have an outlet, so driving down one of these streets to the end and turning around will allow you to drive by anyone who had been following you. This can really make a surveillant uncomfortable and allows you to actually look into their vehicles and possible take pictures. This only works if the surveillance does not know it is a CUL-DE-SAC. If they do they may just stop and wait for you to come back out, or send in someone on foot or on a bicycle to see which house you visit.

COUNTRY ROADS are tricky for a surveillance team because there are few places to blend in. They are usually long with few obstructions and some only lead to one destination. Since the client already knows where he is going and why, the surveillance team will need to make adjustments and attempt to have a purpose for being on that route at any stops and at the final destination. There are several things you can do. Just pull over for a rest break. If any cars behind don't pass you but stop, slow down or pull over that would be very abnormal. Varying your speed as discussed earlier really works well on country roads. If the road has any blind curves or elevated positions you must pass that will block you from the view of anyone behind, you can speed way up, get a lot of distance then slow down, if someone behind you speeds up once they see what you have done then slows down so as not to pass they are staying back there for a reason.

When you are conducting surveillance detection you will probably not need to inform the clients as there will be no extreme changes in the convoy tactics. But if you are going to conduct any anti-surveillance maneuvers you should tell the clients what is taking place. You don't what the client to panic or become upset. Your first job is always to ensure the client is safe; detecting surveillance is a secondary function and you should not do anything that will endanger the clients at anytime. Since everyone associated with the client will be under surveillance, if the client is a target so are your security detail, equipment convoys, and office help. They should all know how to conduct surveillance detection.

In some places you need to remain aware of the local legal consequences of some of the anti-surveillance maneuvers you attempt. You should not be worried about getting a ticket but possibly involving local law enforcement and embarrassing your client and getting his name or name of the company in the local media. Make sure any anti-surveillance maneuvers or surveillance detection is planned out like any other part of the movement. After all, you don't want to get stuck in a cul-de-sac with no way out and several vehicles belonging to the threat blocking your exit.

COUNTER SURVEILLANCE

COUNTER SURVEILLANCE, a strategy used in a security plan that is part of your surveillance detection program, is when you have someone watching you to see if surveillance is watching you, or someone watching for watchers. Counter surveillance (CS) can be done several different ways and should fit into your overall surveillance detection program. It is a proactive part of a good security program for an individual and that company's fixed sites. It will cost more money but that is a decision the client has to make.

First, for mobile counter surveillance you basically already have a CS team in place. These are your low profile security teams that you send out: the security advance patrol (SAP) or the counter assault team (CAT). Since both of these teams are traveling in front of or behind the main body detail, they should be looking for signs of surveillance. The SAP should try to notice anything unusual as they pass by: people putting vehicles in place, taking positions, or coordinating outside. Notes, photos and videos should be taken and passed back to the main body convoy before they depart.

The CAT on the other hand will have a much easier time spotting surveillance since they are passing through the area after the main body convoy passes. They have the chance to catch surveillance teams pulling out behind the main body, surveillance team members leaving the roles that helped them blend in to the operational environment and hurrying away or basically any activity that coincides with the passage of the main body convoy. Once again notes, photos, and videos should be taken and compared after the mission ends to try to come up a plan after identifying surveillance personnel tactics, patterns and people.

FIXED POINT COUNTER SURVEILLANCE is used when you have certain known locations that could come under surveillance or certain routes you must use on a regular basis with choke points you must pass through or other danger areas. This is also more of a long term plan for these type of operations. Basically what you do is establish a counter surveillance point in an area that a threat must use to conduct surveillance against you at a fixed location. You first study the area and determine those locations a surveillance team would have to be located at in order to conduct surveillance against the site or anyone leaving the site, then you find a location that allows you to observe those potential surveillance locations and emplace a counter surveillance team. The CS team will watch for any new personnel in the area, unusual activities and any activities that seem to be triggered by the client's movement.

When you use a CS team at choke points or danger areas you will do the same thing. In these locations there will be certain places the threat must be located so they can conduct surveillance against you or to conduct an attack. You need to identify these areas and put a counter surveillance team in place to observe those areas for surveillance or attack site preparation by the threat. Your counter surveillance is going to have to be covert and blend in with the area. They might have a motel room where they can observe from, a diner they eat at all the time or work at, etc. They need to be in place and observing 24/7 if you have the resources. If not they should be in place at least 3 hours prior to any movement from the departure point.

What should you or the counter surveillance teams do when surveillance against you is confirmed? Well, there are several things that need to be done when surveillance is detected. It needs to be reported to the security detail team leader so a decision can be made to continue the mission or not. This information should also raise the awareness levels of the traveling members of the detail since they know they now have an active surveillance threat against them. Record all of the details of the sighting of the surveillance, include photographs and videos if you can get them. This record and accompanying documents, photos and video are important because they will allow the rest of the security detail to see the people conducting surveillance to allow for recognition if they are seen elsewhere and it will give an insight into their tactics and techniques so they will know what to look for in the future. All information needs to be considered. Now that you know you are the target of surveillance you can adjust your tactics as necessary. Let them see how hard a target you are and they may move on to another softer target.

At the completion of each mission, there should be a meeting of the security detail to include the security advance patrol and/or the counter attack team. This meeting is going to be a quick rundown on what was seen on the first leg of the trip, abnormal activities, people or vehicles that stood out along with any photos or video taken. This is a quick compilation of information before the detail moves on to the next leg of the mission. Review all the information that was collected not only from the team but from any counter surveillance, from your home base or departure point, from the military or local law enforcement on any new incidents that day and your travel route status. You need to go through all the information to put it into categories, then pull out the stuff that will affect you and the mission. You then have to assess that information to see how it will affect your continued mission. Next, you brief the security detail and your security operations center, come up with a new plan if one is needed after taking into account any new information, then continue the mission or don't continue, use your planned route, use all alternate routes, go to a more aggressive security posture or go low profile.

The important thing to remember is your primary mission which is to keep the client safe. Do not spook or confront anyone you suspect of carrying out surveillance. This could lead to an immediate attack, or worse it could be local law enforcement conducting surveillance and you could be arrested. Also don't become paranoid or panic and decide everything is a threat because this will inhibit your ability to make sound security judgments. Lastly, don't change anything with out talking to the client. Remember, you WORK FOR HIM, not the other way around. If you wish to delay or change or cancel something the final call belongs to the client. Make him or her aware of all the options and consequences to help make a decision but it's still their decision.

What will make the threat look for a softer target when conducting surveillance on you or the client is YOUR professionalism, YOUR tactics, YOUR unpredictability, YOUR level of alertness and YOUR rehearsals and immediate action drills. The threat will pick who to attack based on his criteria but part of that decision is going to be based on YOUR actions or inactions.

SETTING UP COUNTER SURVEILLANCE

As discussed previously counter surveillance is nothing more than getting someone to assist you in detecting whether you are under surveillance. Sounds pretty simple but placing in a good fixed site counter surveillance point can be a lot of work. You need to plan a route, which is sometimes called a surveillance detection route. This is to your benefit since you are going to pick a route that you are familiar with, where you know the environment, the demographics, etc. You should have a feel for this route that will allow you to know when things are not normal and when new people are about.

You should conduct an assessment of the capabilities of the threat against you as it pertains to surveillance: how they conducted it in the past and how they use technology. Know their capabilities.

Do a map recon of the routes you want to use. Avoid areas where you will have little or no control of the environment and any other areas that will cause you undue problems in case of an incident.

When looking for places to set up your fixed site counter surveillance it has to be an area that any

threat surveillance will have to follow you through so they can be seen. It also must be an area where your counter surveillance personnel will blend in, fit into the environment and become part of its daily routine so they don't stand out and get discovered by the surveillance. It is also important to set the fixed site counter surveillance up in such a way that any surveillance will be funneled by them. Try to position them indoors in a restaurant, bar, café, or bookstore where they can see the convoy as it approaches and as it departs. They also have to be able to see as many entry points to the route and exit points as possible so that can determine if potential threat surveillance had pulled off prior to reaching the detection point or they can see if another surveillance team is picking up the convoy or the main body. To cover all this you need at least two fixed site counter surveillance points.

When selecting your routes and locations for fixed site counter surveillance be sure to avoid any areas, roads, or other locations that will give alternatives to the hostile surveillance teams so they can avoid the detection areas.

There is another type of counter surveillance that is used when you have to do a movement by foot, and though there are a lot of similarities to a vehicle route, for a foot route there will be some differences. A foot route is usually put together so you have the same two counter surveillance people checking you for threat surveillance along the entire route. This way they do not need a video to search for any similarities because they will recognize the same person or car immediately. To accomplish this they have to be able to move from one fixed point counter surveillance location to the next site on the route. So while the detail is walking at a certain pace, the route is usually meandering or they will make planned stops to give the fixed site mobile surveillance time to pack up and move to the next position.

I prefer having more permanent type fixed site counter surveillance and people for longer periods, especially around the fixed sites of the clients or other facilities you will be working from. Hiring vetted or trusted locals or expats that have the ability to speak the language and blend into the area for the long term is a great way to set up a surveillance detection program. With the modern technology now available, real time video, wireless cameras, cell phones, and palm pilots are useful. A good fixed site counter surveillance program will greatly enhance your overall non-visible security posture.

If you never detect any surveillance the expense, time and effort in setting up a surveillance detection program was worth it since confirming you are either under hostile surveillance or not under hostile surveillance will determine how you proceed in almost everything the security detail needs to do when conducting operations.

CHAPTER 10
AWARENESS LEVELS FOR THE SECURITY PROFESSIONAL

COL Jeff Cooper, a former United States Marine who saw service in WWII, Korea and Vietnam, went on to become one of the finest combat shooting instructors in the world. He developed a color code that allows you to track your four levels of awareness depending on your training, experience and how perceptive you are to the operational environment.

WHITE: This is the first stage of awareness, where most inexperienced untrained people are most of the time. It is a state of total unawareness in your environment. You are daydreaming, watching a pretty girl, a very nice car, you hear a song on the radio and are humming/singing along. When people are in this state they have a hard time noticing any type of indictor they observe. It's like being in your own world and while your eyes are scanning, you are not actually seeing anything. This level of awareness is also the result of being fatigued, consuming too much alcohol the night before a mission, etc.

YELLOW: This is the next stage, where you are relaxed, but cautious. It is where most security people are doing day-to-day tasks when the threat level is at the normal stage for the client and there is no specific threat anticipated. You are not distracted and are paying attention to what you are doing. You notice indicators when they are seen, and do not get tunnel vision on a specific thing but keep scanning your areas of observation. This is where a trained and experienced security detail member will be at most of the time on a mission.

ORANGE: When an indicator is observed and a potential problem exits your awareness level will move into this stage. This is a specific state of alertness for a specific threat you have perceived. Something has stood out and takes your full attention as you examine it further. You start thinking about how to engage the threat and ways to neutralize the threat or avoid it alltogether. You also warn the rest of the detail of the threat. YOU MUST NOT GET TUNNEL VISION on this potential threat because it could be a diversion.

RED: There is no way to avoid the threat at this time. You cannot bypass it or turn around. The engagement with the threat is about to begin, and you and the team have to be ready in some way. You have thought through all your contingency plans and SOPs, and have come up with the best course of action that you and the team can think of to protect the client and you put that plan into action.

Some people in the security field have little or no training and no experience which can be a big detriment to surviving an encounter with the threat. Those untrained, inexperienced people usually end up in a state of pure panic where they freeze up and can't do anything, or the other extreme where they empty their weapon into their car door. The whole point of any ambush is to achieve total surprise and to induce panic in victims, and other people in the area. The threat wants to create sensory overload, so you need too keep an eye on your teammates. If they freeze up, give them some direction and hopefully instinct should take over. If they start panic fire try to make sure they are not hurting innocents, team members or the client.

The role of a PSD is to protect the client. This is done by anticipating what the threat can and will do and preventing them from being able to do it. PSDs that rely only on reacting to a threat incident have already lost control of their operating environment. The PSD must reduce the chance of an threat incident in such a way that the potential threat will be detected before an incident occurs, or while the threat is conducting their surveillance so they will decide you are too hard a target and move on to a softer one. Whatever the reason, no incident that happens is a success for the PSD. Knowing when to be prepared, when to raise your awareness levels will come with EXPERIENCE, training, knowing your duties, EXPERIENCE, observation skills, and understanding what you see. These are much better skills to have than how fast you can run, how much you can bench press, how much cool guy stuff you have and how big your gun is.

CHAPTER 11
SITUATIONAL AWARENESS

The role of a private security company and the private security detail is to protect the client. The best way to achieve this is to avoid any attacks or other hostile situations wherever possible. To accomplish this, the security detail members must be able to recognize any danger signs or warning indicators that are present. Every time you have to **REACT** to a hostile situation or event you have lost the security initiative and are no longer controlling the situation. You will have to fight to regain this control in order to protect the client. The key to situational awareness is to recognize the various danger signs and indicators that are presented before an attack or when other hostile actions take place. If you can recognize these signs before the attack and evade or counter the hostile action before it starts then you have maintained control of the situation.

Another trait that needs to be developed by the individual security provider regardless of the type of security detail is to become third party aware. Many times a protection detail will focus on the clients and those people moving with the clients which causes them to lose awareness of their surroundings. This is important because in every attack that takes place there will be warning signs or indicators that need to be picked up by the security detail prior to the attack taking place. If not, then the detail will be reacting to the threat action, not acting before the threat materializes and when you are reacting you are usually too late to stop an incident from occurring.

Your **OBSERVATION** and subjective interpretation of what you see influences how you react to any situation, and what tactic, SOP, or drill you decide to use. What you decide to do will then influence how aggressive and more importantly how accurate your action is so the earlier you observe an action and have a clear understanding of what is taking place, the advantage switches from the attacker to you, as you start countering the threat they represent before they are fully ready to complete their action or are capable of countering your action.

Observation actually has 3 main elements as it pertains to this subject:

1. **ATTENTION**: When working you are always alert although your levels of alertness vary depending on the operational environment you are moving through (awareness level color code). When you are working, you are scanning in your area of observation and it is important that you actually SEE what you are looking at when scanning instead of just looking in that direction. Something is always trying to draw your attention/eyes in several different ways:

- **Involuntarily** drawing your attention happens when there is a sudden movement, loud noise, loud colors, bright light, or flash of light. These types of things usually draw our eyes to the object or person that is creating this phenomenon. Sometimes the threat uses these types of actions or sounds to create distractions to draw our eyes away from the true incident forming, so do not let your involuntary actions control your attention.

- **Voluntarily** focusing your attention is what you do when you are scanning your designated area of observation, deliberately looking into specific area or at certain objects. Your scanning ability is something that improves the more you do it. When scanning you should break your area down into three areas of observation, the foreground, middle distance and far distance. You need to sweep your eyes slowly back and forth in each of these zones concentrating on anything that can be made out. Don't focus too long on any one object or person, mark it, continue your scan and come back to it.

- **Habitual** attention is a phenomenon that happens when you get so used to seeing something that when it changes, you can't quite figure out what is wrong or what has changed. Your brain can also play other tricks on you when you have something you are used to seeing. Say you are always used to seeing a policeman at an intersection and one day you see him there but you get the feeling something isn't right. You look at the video over and over and can't figure it out when someone else can come in and look at it and tell you that his uniform is different with one look because he wasn't used to seeing it the way you were so he could tell the differences much easier.

2. **PERCEPTION** is your understanding of what you see. This in part is influenced by your mission, your experience and your working environment. Since a large part of perception is based upon personal experience, people rarely perceive things the same way, so you need to discuss what you see with other members of the detail to get their perceptions of the situation area.

3. Next is **INTERPRETATION**. You need to interpret what you are observing for possible action and you do this by first deciding exactly what it is you see (analyze), and remembering that the threat, whether hostile or criminal, might be trying to deceive you into seeing something harmless, non-threatening. Once you decide what it is you see, you have to figure out how it is going to affect you and what you plan on doing. You must quickly decide on what course of action to take, then DO IT. This is why we plan, rehearse and "what if" scenarios constantly. A quick decision might not always be the best decision possible but hesitating or making no decision can get you and the client killed.

CONCLUSION: Once you believe you have a grasp of what's going on and understand what you see, you need to form a conclusion and decide what course of action, if any is necessary, will best suit your needs of stopping the perceived threat. All this will happen very fast, that's why training, experience, and rehearsals are so important.

Learning to recognize any danger signs or indicators is very important since they will help you understand what you are seeing and what appropriate actions to take. Danger signs or indicators can be broken down into three main categories:

1. **OBVIOUS SIGNS** is just what it sounds like. It could be someone holding a weapon, pointing an RPG at your vehicle, a street blocked off with burning tires, a mob of rioters moving towards the facility, or a truck full of armed men stopping outside a gate.

2. **SITUATIONAL SIGNS** are indicators that on their own would be nothing but when you factor in your situation at the time that these indicators are present it could be possible signs of trouble, such as an unofficial check point passing all locals through it, children who normally sell newspapers and drinks along a route are missing, coffee kiosks or cafes along the route are closed and the furniture is taken in, or a vehicle blocking a road with lots of people around it not trying to fix it.

3. **BEHAVIORAL SIGNS** or indicators are those things you can physically see when looking at a person who is a potential threat. These indicators can be divided into three subcategories:

- **SIGNS OF INTENTION**: A person displays many types of physical indicators when they are preparing to become aggressive or violent. In the face you will see a fixed stare, clenched teeth, lowering of the eyebrows, squinting, and thinned lips. There are also indicators with the body when someone is about to get aggressive or violent, including assuming a stance to fight, clenching their fists, staring at you, hunching the shoulders, and a change in breathing. Finally, its their actions that are indicators, how they are moving towards you and the client, walking aggressively or hesitantly, etc.

- **PHYSIOLOGICAL INDICATORS**: The body does many things that are beyond a person's ability to control. One of them is the fight or flight reflex. This is an automatic response to threat or extreme fear or excitement. A person in this state will get very pale as his body reroutes his blood to his vital organs and major muscles for fighting. That blood is taken away from the outer vessels which is what leads to a paling of skin. Heavy breathing is another indicator that can easily be seen and is very difficult to conceal. The body is preparing itself by oxygenating the blood and getting ready for action. He will have a dry mouth as the saliva is deceased, which can cause a person to continually lick his lips. Muscles tense up prior to any action because of the increased blood to those muscles in preparation for action, so a person might clasp his hands together tightly, cross his legs firmly or have his arms crossed hard across the chest. And because all this new activity is taking place, the body is heating up and will attempt to cool itself by sweating for no apparent reason.

- **MASKING ACTIVITIES**: Surveillance personnel and attackers must find a way to blend into the environment they are working or waiting in. As we know most surveillants have a fear of being detected but so do potential attackers because this is the riskiest time for them. Not only are they going to attack armed men, but they have to worry about being caught by local law enforcement or military with weapons or other items they are probably not supposed to have which can cause paranoia to set in so attackers tend to over compensate trying to act normal. This can be compared to waiting for a pot of water to boil. Time seems to stand still, and to fill in what they perceive to be a long time, they begin fidgeting, performing repetitive activities, looking at their watch, playing with a ring or their hair, fixing their clothes constantly, picking up something to read then putting it down then picking it back up. Other indicators are half finished cigarettes in ashtrays or on the ground, unfinished drinks such as coffee or tea, not getting refills on empty drinks while waiting, playing with the glass, etc. The best way to identify this behavior is to have fixed point counter surveillance in place at choke points and at potential attack sites along your route. If you do not have the resources for this, the security advance patrol can stop at each potential attack site or choke point and observe the route prior to the convoy passing through. Since they are low profile they should be able to blend into the local environment. This is also why it is important to vary times and speeds of vehicles during transport. If a threat is expecting you to pass an area at 1000 and you pass at 1030, the threat will start becoming agitated because of the increased amount of time that passes after you do not show up at the time expected. He'll start thinking that you are not coming or something could be wrong, so the indicators will become more pronounced.

When talking about indicators of a possible threat you also have to look for people who are transitioning from the non-violent to the violent, a non-threat to a threat. In order to carry out an attack against you the attacker will need to switch from the cover for blending in to a more suitable posture for the attack. This can include standing up and turning towards you as you pass by, putting out a cigarette as you approach, slamming down a drink as you approach, reaching into a bag that is sitting at his feet, reaching into his coat, moving towards you from his previous position, moving against the crowd or pushing people. There are many things a person has to do in order for him to get into position to carry out the attack, and you have to be able to recognize when this is taking place. All of these indicators are visible and hard to disguise as long as you understand what you are looking at when it is happening.

TACTICAL MINDSET

There is always a need to get into that tactical mindset when conducting operations, which goes hand in hand with your situational awareness and perception of your operating environment. You should always be analyzing your situation, deciding on the best course of action to take in case of an incident so you will always be in control of the action and situation. Everyone should do this by conducting what is called a moving threat assessment. As you approach a possible problem area such as a choke point or other potential danger areas or attack sites, look at the site and situation as you approach. Decide what tactical plan would allow you to establish control in case of an incident and allow you to neutralize any threats if they materialize.

Each detail member should have his area of responsibility already assigned; this will be his area of focus during movement and in the areas that require heightened awareness. As you pass through these areas or into other potential threatening situations you should be thinking, "what are the potential threats?", "do I now control these threats?", if not "how do I control these potential threats, if necessary?" Constantly run attack scenarios in your mind and develop solutions for them in your area of responsibility.

When you have identified a threat in your area of responsibility, this will become your main area of focus, and you will implement your plan to neutralize this threat. With each member of the detail performing this function you maintain overlapping areas of protection, which will allow any threat to be recognized and neutralized as quickly as possible. Constant training, experience and communications will increase your situational awareness to the point it will become second nature, improving your and your client's chances for surviving any incident.

CHAPTER 12
GENERAL GUIDELINES
FOR PSD OPERATIONS

• The protective detail must always provide and maintain a 360 degree security bubble around the client.

• The type of security detail provided and the number of personnel and vehicles in the detail will depend on the mission, the threat, and the availability of manpower and resources.

• Members of the protective detail, either in the vehicle or during foot movement, should not be looking at the principal or at distractions in the vehicle but should always be facing outwards.

• Protective detail members should be looking at peoples' hands when viewing personnel in crowds during foot movement and during vehicle movement. When a person transitions from the cover he had to adopt in order to blend into the operating environment to a THREAT what the hands are doing is usually the first indicator.

• When you are in tight crowds either during foot movement or vehicle movement you must have tight coverage. When the crowd thins out you can expand your coverage out while in these open areas.

• A private security detail's job is to protect and then evacuate the principal away from a threat incident. It is not to engage or pursue these threats. This is something many security details forget, the cowboy in them takes over. I have read incident reports where security companies have actually left the clients in their vehicle by themselves so the driver and detail leader could get out to engage a threat because the scout vehicle and gunship both stopped to engage the attackers instead of pushing through. DO YOUR JOB: get the client out of danger to a safe area!

NOTES

NOTES

CHAPTER 13
FOOT MOVEMENT TACTICS AND TECHNIQUES

Moving by foot can be more dangerous since the client is outside the protection of a vehicle or building, and the security has more distractions to deal with. It is important to practice this type of movement so everyone understands where they are in the formation and what their responsibilities are. Below is some basic information on foot formation tactics.

The **type of formation** you use will be dependant on several factors: the available manpower, available resources, degree of threat for that event, and other security at the event. The greater the threat for the event, the tighter the security sphere should be around the client and/or clients. The more spread out the crowd the more open the detail can be; the closer or tighter the crowd, the tighter the security detail has to be. In the event of an incident, the nearest detail member responds to it. The other detail members close up the bubble putting that security back into place, and they move the client away from the incident. The detail leader or bodyguard will physically lower the client's profile to get him totally behind the wall of flesh that is the security detail and he will direct and cover the client until the incident is over and during movement to a safe area.

In **large crowds** with no other crowd control, the detail will tighten the security sphere around the client, keeping people away. Anything that is thrown toward the detail will be struck away from the detail and to the ground immediately, since ground blasts are less damaging then air bursts. The thing to keep in mind is any formation you use needs to be flexible and have the ability to adapt to the operating environment and to

the needs of the client.

There may be **Special Situations** that need to be considered at some locations when protecting a client including:

- Receiving lines
- Revolving doors
- Stairways
- Escalators
- Crowds
- Press

There are some additional security measures that can be used in **high density crowds** or venues when there is a high risk factor and the resources and personnel are available. You can have mobile crowd watchers, which are low profile security personnel who blend in with the crowd and move through it, pacing the client on the left, right, front or rear. They are looking for people who are keying on the movement of the client, who themselves may be pacing them and for anyone who may be transitioning from a normal event observer to a threat to the client. Static crowd watchers are people who are low profile, blending in with the crowd, while taking up stationary positions where they can observe the people around the client and the security detail as they move towards the client's destination. You can secure all the vantage points of the venue. If the site manager allows it this will give you a greater level of observation and increase your tactical response in case of an incident. You can use counter snipers at some sites as long as you have trained personnel. This will give another layer of security as well as greater observation of the area as your client arrives and departs the site being visited. You can have a counter assault team on standby within the site itself located in an area like the safe room, in a building across the street, or staged in a vehicle parked close to the site.

ONE MAN DETAIL

• Simplest formation, used when the whole detail cannot accompany the client into a facility or for an event. Also used in low threat environments and situations.

• Detail leader or bodyguard, whichever has been assigned for this event, will have 360 degree coverage of the security sphere by himself.

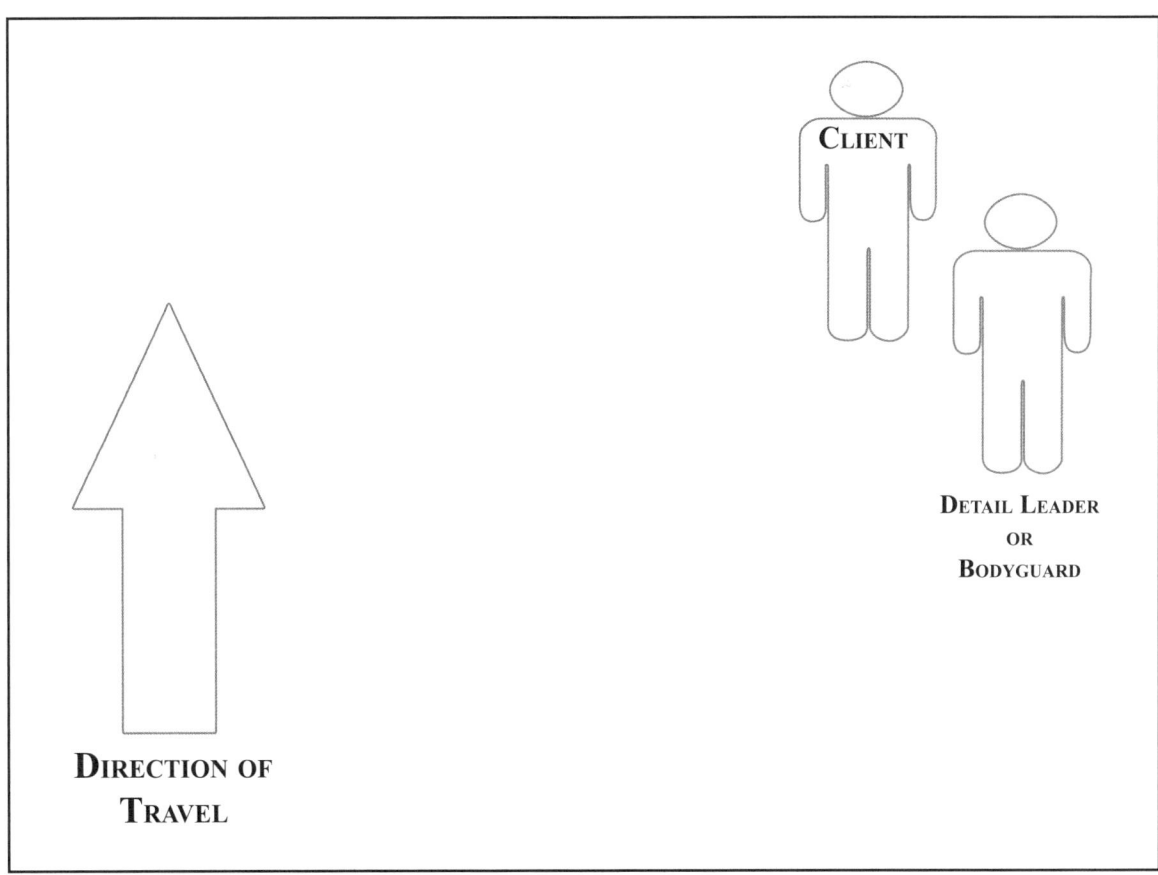

TWO MAN FORMATION

The detail leader or bodyguard will share the coverage of the client with the advance leader; each will take 180 degrees of the coverage. This is used with in a low threat environment and when you have an advance detail. Since the advance leader knows where the client has to go and how to get there, he leads the way.

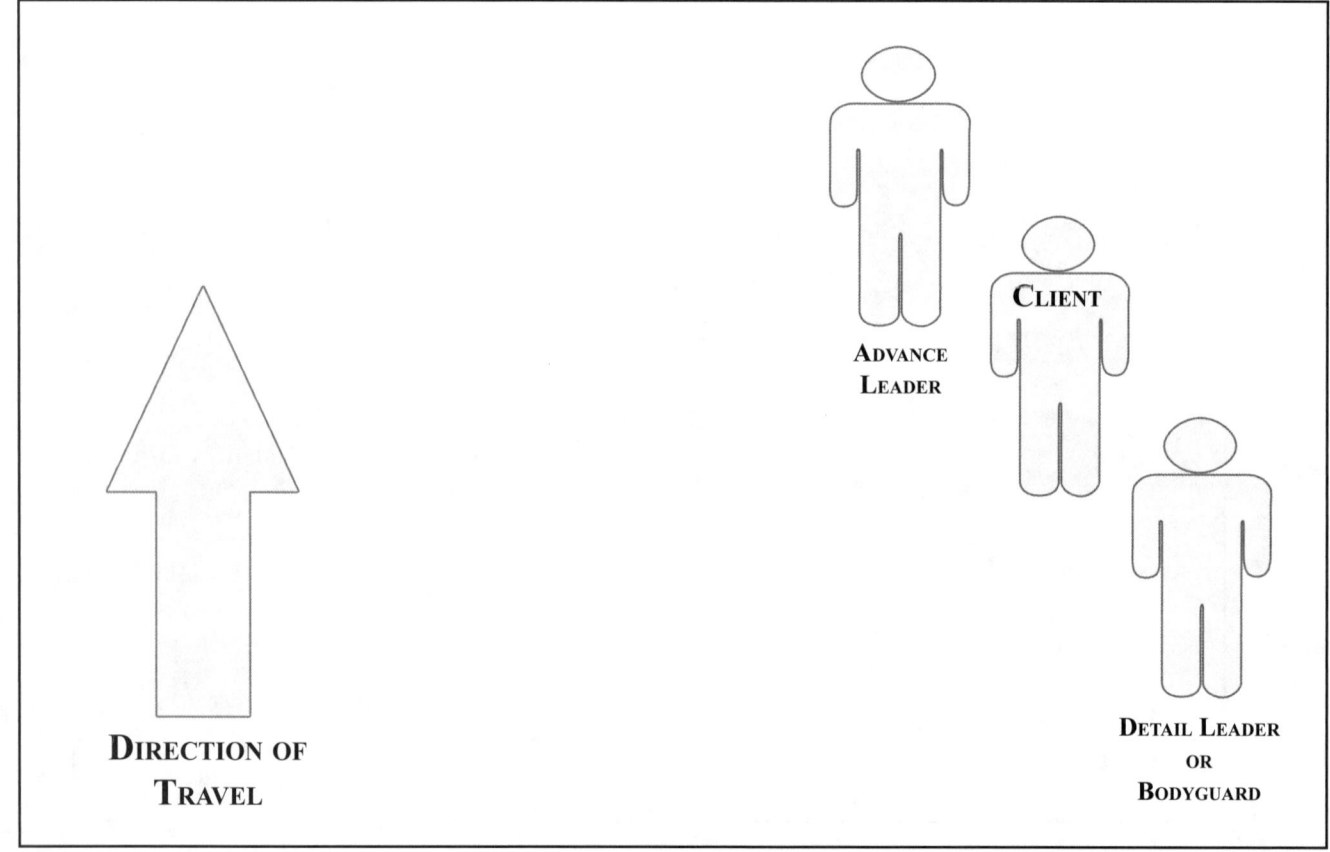

5 MAN FORMATION
(MODIFIED DIAMOND)

The shift leader, advance and two other security team members will set up the security sphere around the client by sharing equal areas of coverage, still providing 360 degrees of observation. The detail leader/bodyguard will move inside the security sphere to the rear of the client and is now directly responsible for him during an incident. The advance leader, when available, will lead the way because he knows the location. Two security personnel are to the left and right, and the shift leader is bringing up the rear.

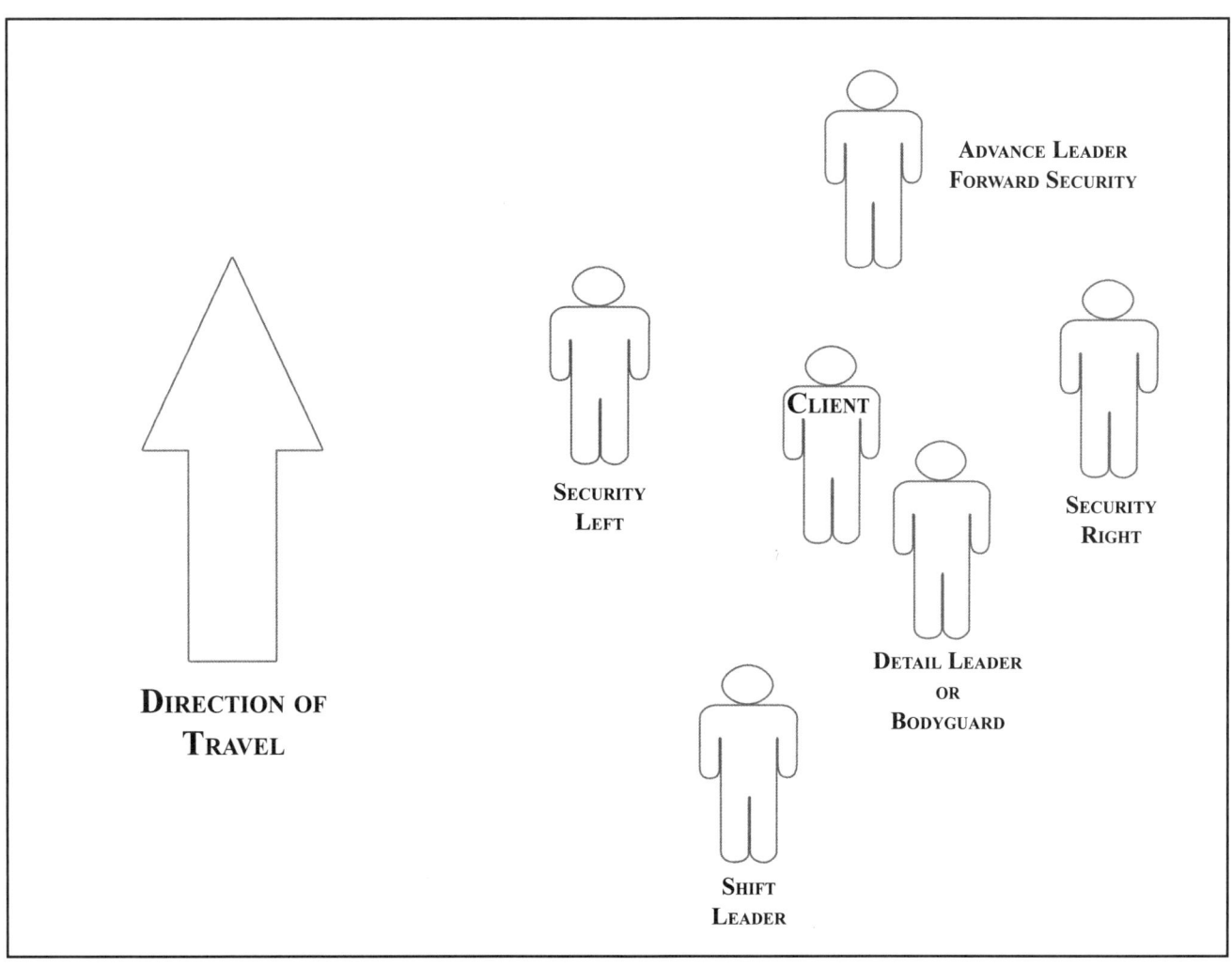

5 MAN FORMATION
(MODIFIED V)

The shift leader, advance and two other security team members will set up the security sphere around the client by sharing equal areas of coverage, still providing 360 degrees of observation. This is more of an open formation used in low threat areas, so that it does not enclose the client, allowing him to seem unprotected while still providing 360 degree coverage. The detail leader/bodyguard will move to the right of the client in the security sphere and is directly responsible for him during an incident. The advance leader, when available, will lead the way because he knows the location. Two security personnel are positioned one to the left and one to the upper right, and the shift leader will bring up the rear.

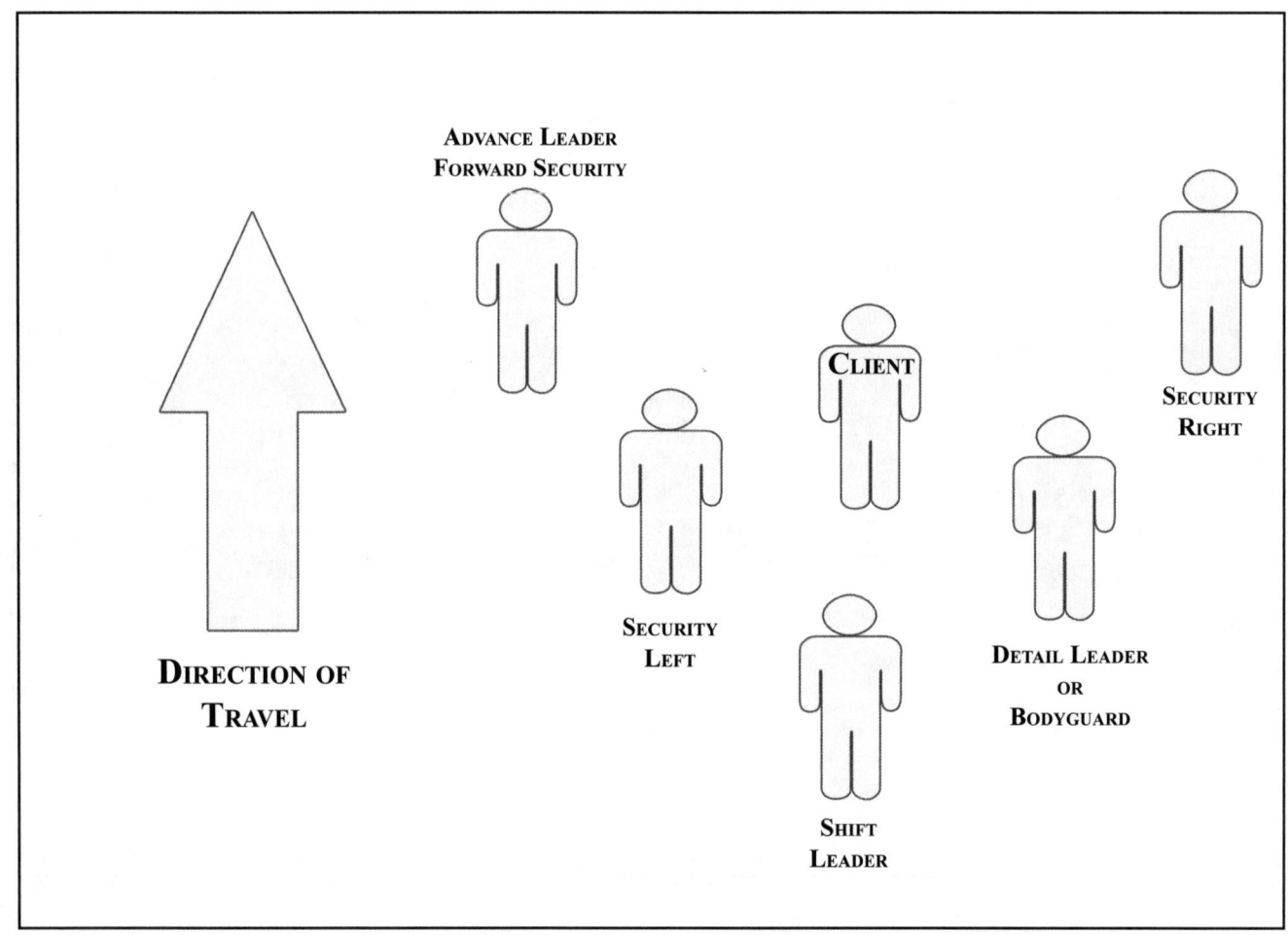

6 MAN FORMATION
(DIAMOND)

This is the most effective formation for foot movement of the client, and normally is used when the security detail has enough manpower and resources.

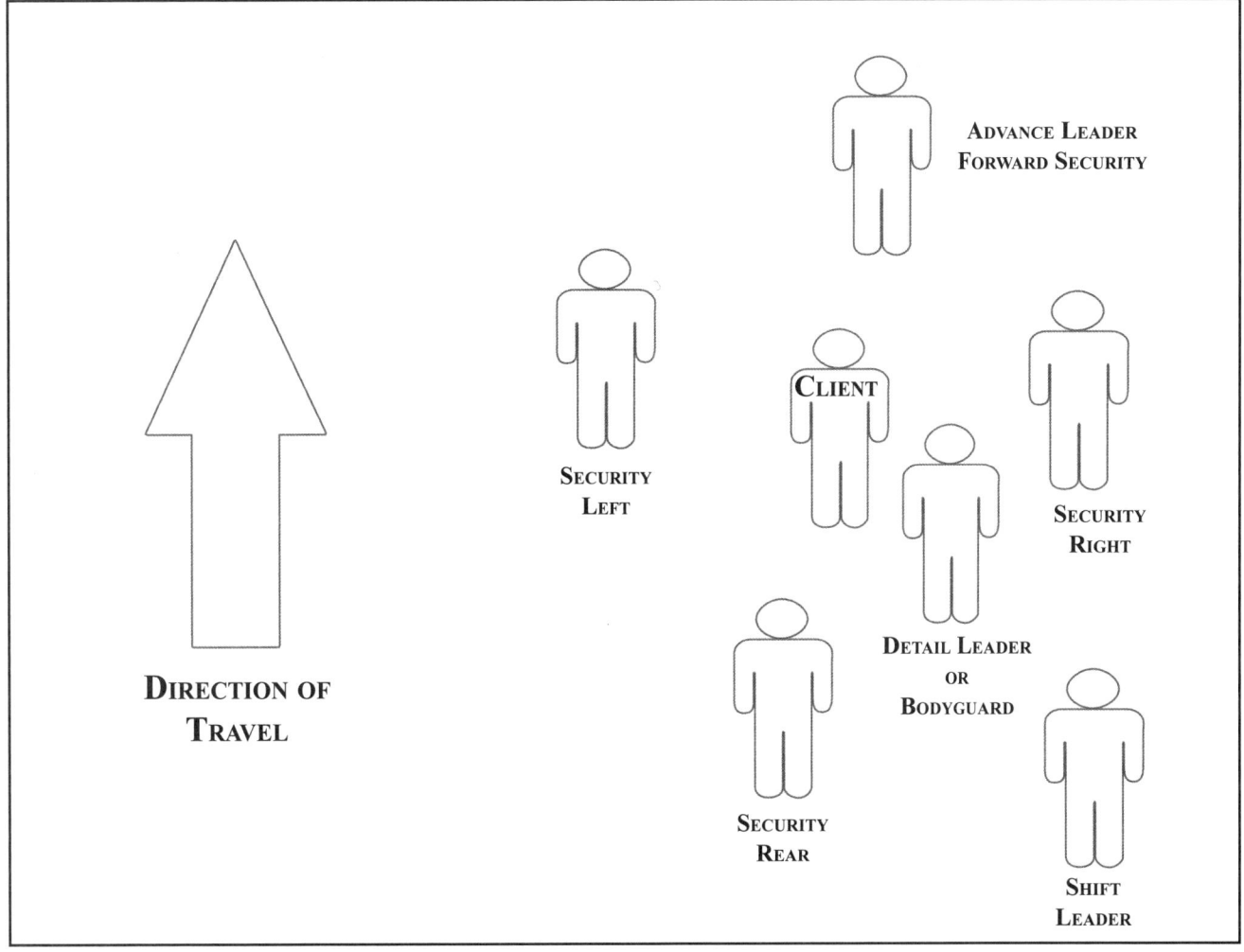

FOOT DETAIL
(OPEN)

This is how the security detail will form when there is a minimal crowd in the operations area and the threat is low.

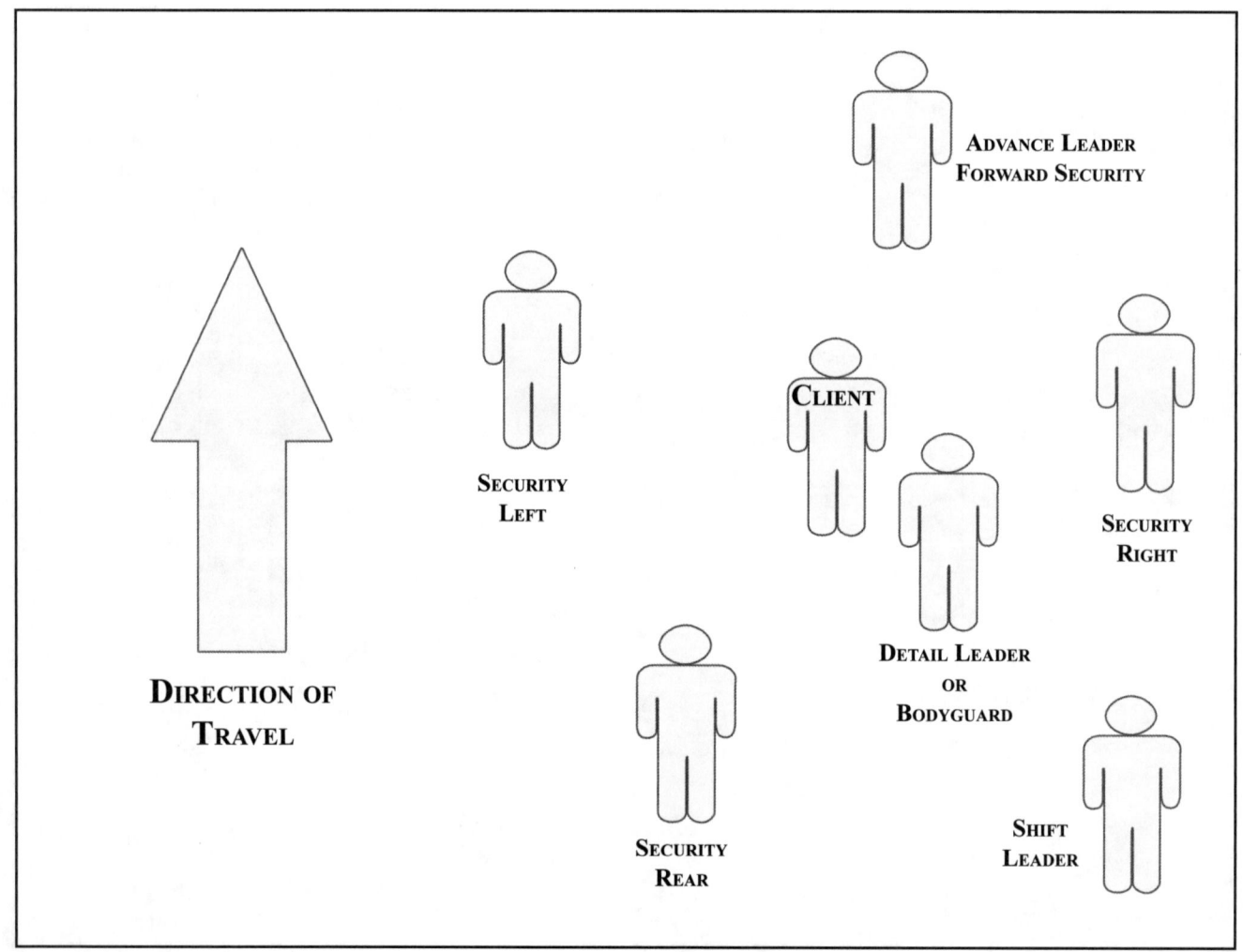

FOOT DETAIL
(CLOSE)

This is how the detail will form on the client when there is a large presence of outside people who are not part of the event. Large crowds, depending on the event, have the chance of turning into bad situations. You will tighten the formation around the client, keeping 360 degree observation and tightening the security bubble around him. You will also use a close formation or tighten the security bubble when the threat is high for the operation.

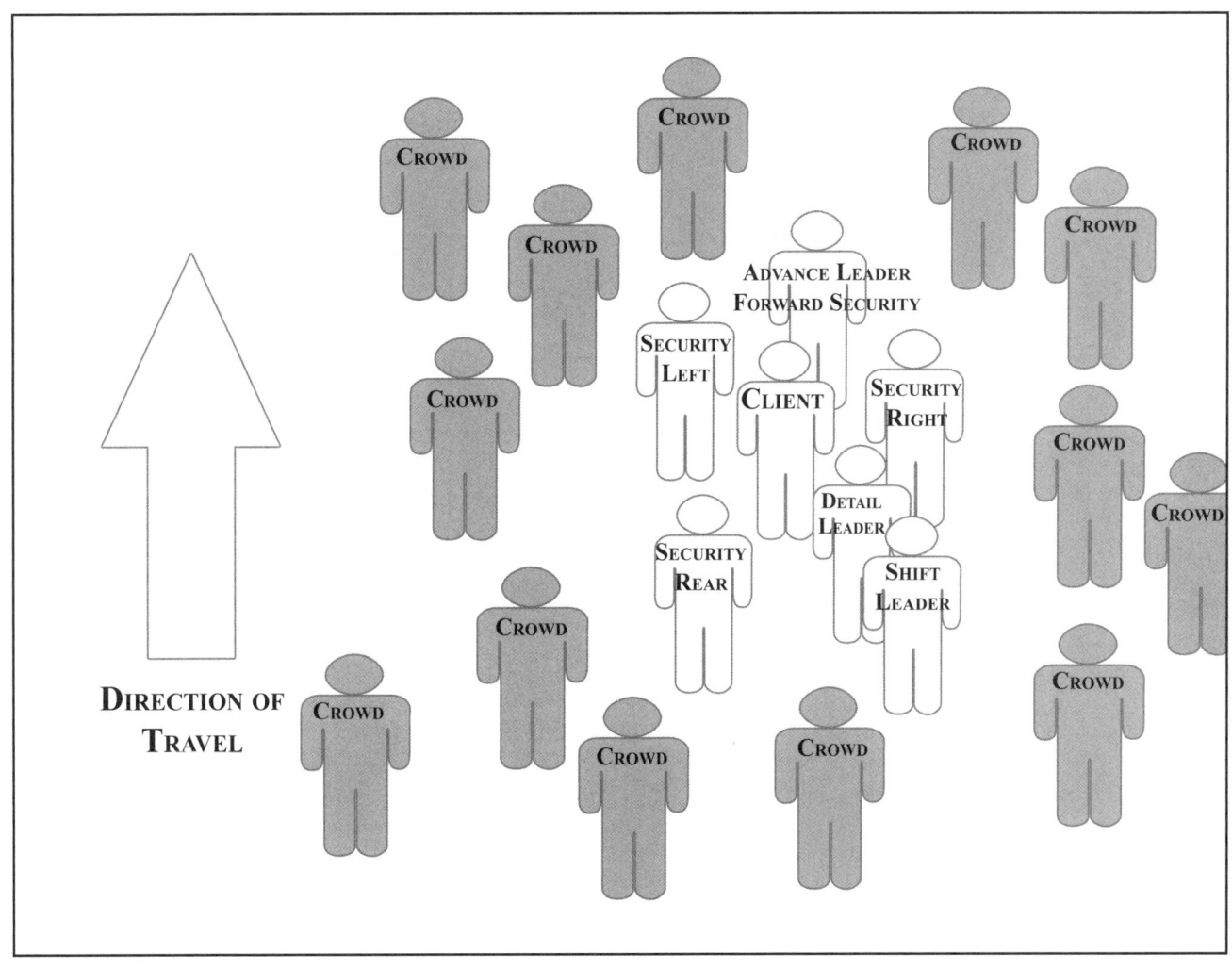

THE SECURITY DETAIL'S MISSION

The first and only mission of any personal security detail is to PROTECT THE CLIENT. In line with this you should also advise the client or clients, and their families on PERSONAL SECURITY so they can start developing that security sense necessary to make them less of a target.

BE PROFESSIONAL

You do not want to embarrass the client for any unnecessary reasons and overreacting is an unnecessary reason. You are supposed to be a professional. (Note: I had a high ranking Army officer and Coalition Provisional Authority member visiting a site that was close to completion. The CPA was a female, who happened to walk around a vehicle into one of the security detail who was taking a piss on the wheel of the main body vehicle with port-a-potties less than 20 feet away. Be professional.)

Also, arguing with the client in front of co-workers or his clients is unprofessional. React appropriately to the situation and make sure all your people do also. The client is the one paying the bills. He is paying you for your service and advice, so you should not contradict, deny, or refuse to do things that are in your contract. You must follow the client's schedule as closely as possible. He has one for a reason. If there are changes that need to be made such as departures or arrivals then discuss it with him, DO NOT ORDER HIM, GIVE ULTIMATUMS OR TRY TO BULLY.

ADOPT A SUITABLE IMAGE

Depending on the type of environment you will be working in, you might need to appear non-threatening both visually and in your demeanor, with your weapons concealed and a relaxed but alert demeanor. In a higher risk environment you might need to have a full combat kit on, and appear threatening or intimidating. You should discuss this with the client before a movement.

COMMUNICATE

Continual and concise communication is important with each member of your security detail and the operations center, but don't forget to include your client. He is the one who sets the tone and the schedule for these events. Be available for him to discuss changes and pass changes and information to him as the detail leader deems necessary.

HAVE THE PROPER MINDSET

You also need to ensure all your security detail have the proper mindset for protection operations. Some people who come to this line of work can never reach the proper mindset because it is not what they thought it would be or what they want to do. What needs to be discussed in rehearsals, briefed to new team members and discussed so they do not forget is that PSDs operations are defensive in nature. This will require great observation and preventive skills as well as quick reaction to incidents when they occur. This takes lots of practices, rehearsals and live missions to become instinct and to have successful operations.

INFORM THE CLIENT OF POTENTIAL RISK

You should also make sure the CLIENT is in the proper mindset. Many times companies hire personal security companies for their people and the people don't feel it is necessary. Make sure the client understands the threat, not just the company VIP but every person in the company you are responsible for! (Another war story: In Mozul, Iraq an engineer for a company that I was working for had been borrowing a vehicle to make a run outside the protected perimeter to a local liquor store since booze was not available on military installations. The security company was letting him go, sometimes sending a driver to take him because he bought them beer also. This was stopped as soon as I got there. The engineer, in a fit, was explaining to me how safe it was. Two days later the market he was using got hit by a VBIED.)

CHAPTER 14
PUBLIC VENUES AND FUNCTIONS

Venues such as large meeting places, theatres, movie houses, school auditoriums, any place with a number of fixed seats and a main stage can be a pain in a high threat environment. There must be an advance to look over the site, know the layout and set up a security observation bubble to wait for the client to show. Once the client arrives he will need to be escorted to his seat. Whenever possible the detail leader will sit on one side with the advance leader on the outside closest to the aisle, since he knows the layout. Security men will be by each exit, with at least one in the lobby and one outside by each possible exit from the facility. In the event of an incident, unless the incident is directed at your client specifically, you should cover him and keep the client in place until the crowd has finished running around blocking the exits. Once the exit is starting to clear, the detail leader will inform the drivers and the entire detail where he wants to exit, then they will proceed with the client to that exit. If you cannot cover the site like this due to manpower shortages, you should at least have the detail leader or bodyguard with the client and one other security detail member in the room to give support as necessary, with the remainder outside.

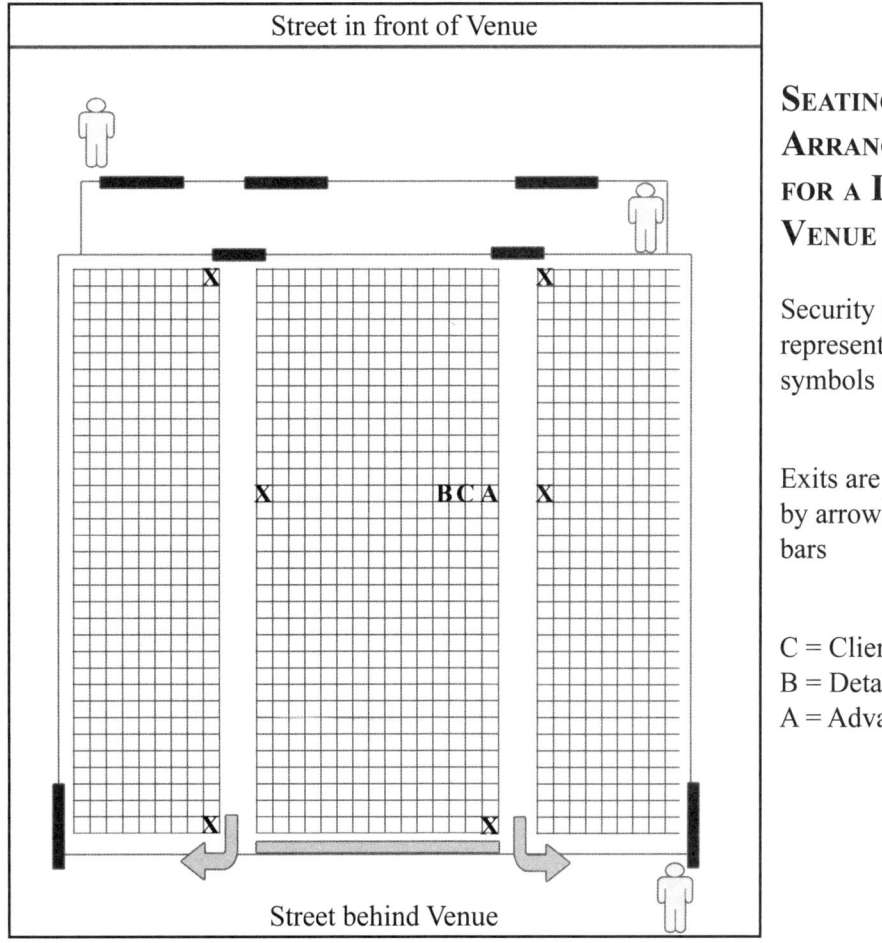

Street in front of Venue

X X

X BCA X

X X

Street behind Venue

SEATING ARRANGEMENT FOR A LARGE VENUE

Security is represented by man symbols and x's

Exits are represented by arrows and black bars

C = Client
B = Detail Leader
A = Advance Leader

PUBLIC SPEAKING VENUES

This is the security set-up when your client is the one who will be in front of the crowd on the stage or raised platform in a large venue.

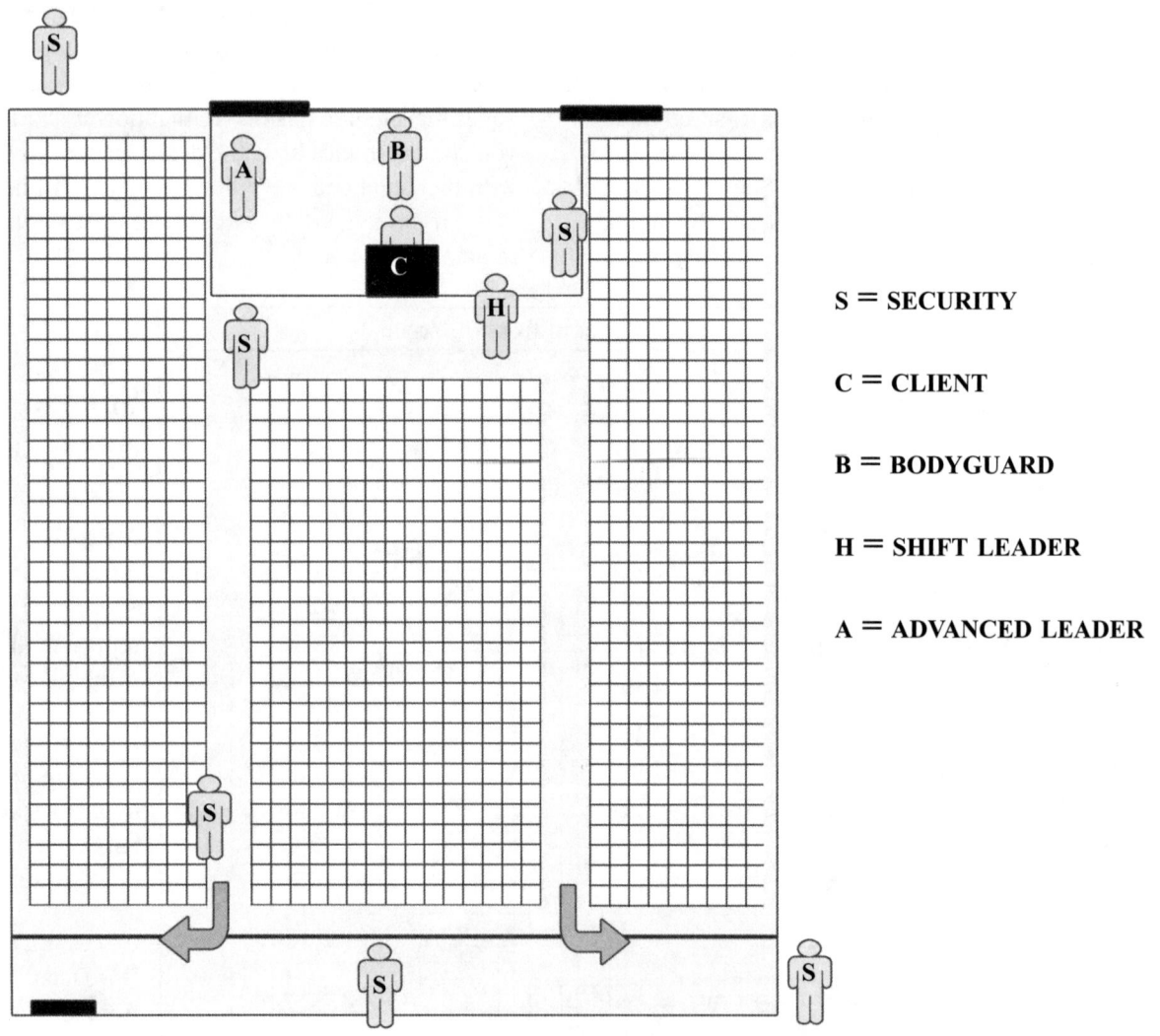

S = SECURITY

C = CLIENT

B = BODYGUARD

H = SHIFT LEADER

A = ADVANCED LEADER

RECEPTION LINES

Receptions are a part of meetings and social gatherings for the client to meet potential business partners or business clients, government leaders or leading citizens in the local community, so the reception line is something that will have to be dealt with. A couple of things to remember: you are the security detail for your client, but that does not mean that you can boss the facility security or security details for other persons attending around!! It is best to work together, which will give you more in-depth coverage for everyone.

When your client is receiving people; if you have the resources and manpower, you need to have 360 degree security along the line and the room. One security guy should be stationed on each side of the entrance way so they can look people over while they are in line and they can clear the doorway if the client has to evacuate or close the door and secure it if there is an incident in the outer layer of security. The advance team leader and detail leader will be behind the client, and the shift leader will be on the opposite side so each person approaching the client will be observed from both sides for suspicious movements. Other security people will be placed around the facility and in the room as the detail leader deems necessary. (See illustration on page 94.)

A receiving line with the client waiting in line to introduce himself and proceed to the event will be similar in some aspects. The detail leader or bodyguard will accompany the client in the line, standing to the rear and to the right. He will not introduce himself or interfere with the other people in the line to meet the hosts. There will be security persons in the room preceding the receiving line by each entrance. The shift leader will bypass the receiving line and stand at the end, ready to lead the client where he needs to go. There will be a security person on each side of the end of the receiving line to provide 360 degree protection for the client. This set-up is for a high threat, high risk event or operating environment, when you are providing the only security. (See illustration on page 95.)

If you are in a permissive low threat environment, this many personnel might not be needed. You can use just the detail leader and the advance leader. You could also be at an event where there is already in place security. In this case the advance needs to coordinate with them to find out what they want them to do so that you can help your client and help in the overall security posture. Remember your client could be the richest most influential man in the room but that does not mean you can do whatever you want as security for your client. Don't be an ass, work with other security elements especially at their home facility. If you have concerns, discuss them with the client, explaining the dangers and give him some courses of action to follow. Let him decide whether to continue the operation or not.

NOTE: I have seen one security element try to rule it over another, making demands at a facility where they were visiting. The site security manager at one location had an advance that arrived at the gate and started demanding stuff. The site security manager escorted them out, and when the detail arrived with the client the site security manager wouldn't let them in. When the client talked to the site manager and found out why he wasn't allowed in, he left his detail and advance on the street and went in by himself because he had to sign papers and get papers signed to get paid for that months work. A PSD is an employee of the company and a visible presence of the company to others. You make suggestions, not decisions on client business matters and your only job is to protect the client. If you get a reputation as a diva company or detail, others will not want to work with you or allow you onto their sites because they don't need the headaches.

SECURITY SET-UP FOR CLIENT IN A RECEIVING LINE

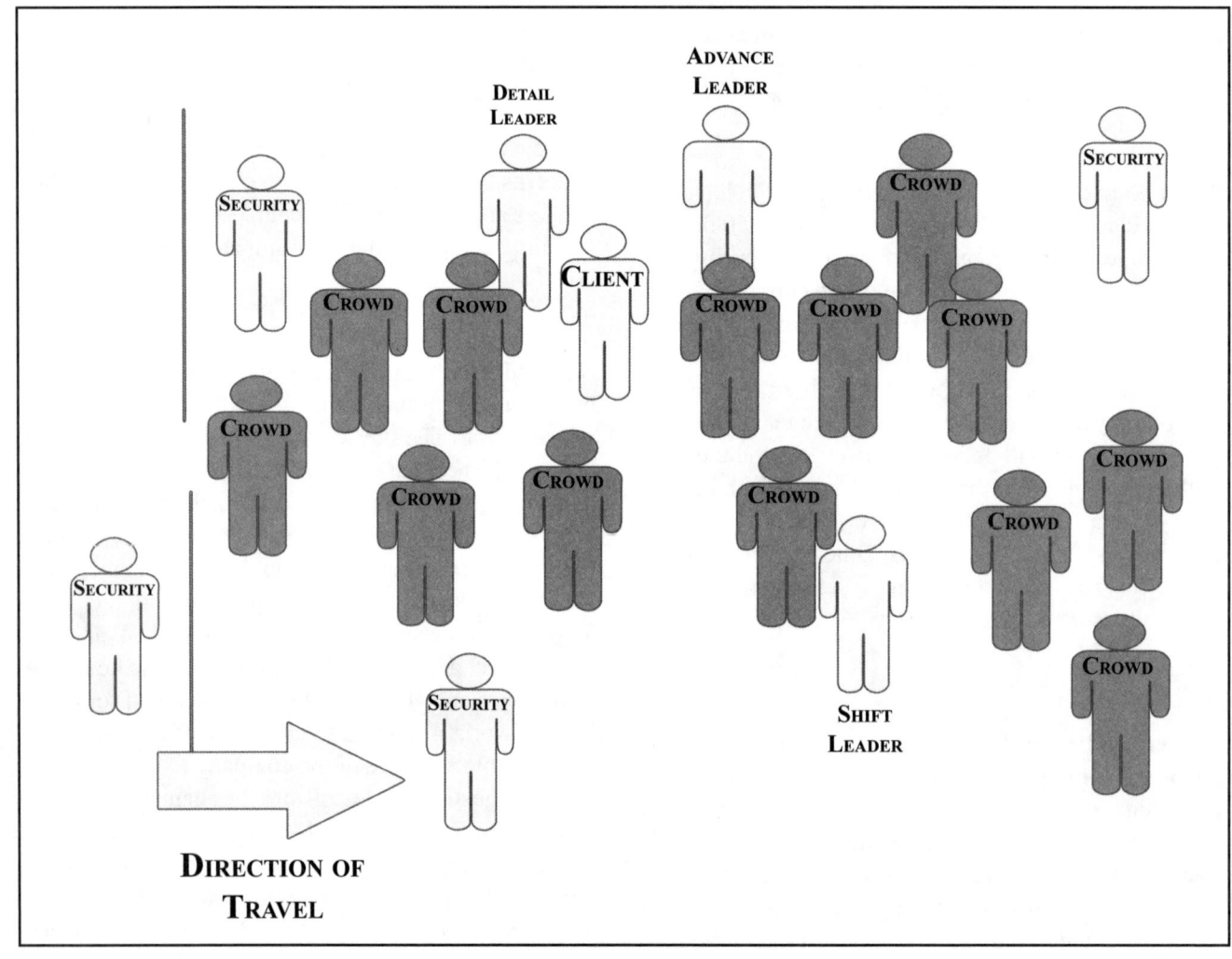

SECURITY SET-UP FOR A RECEIVING LINE WITH CLIENT INTRODUCING HIMSELF TO OTHERS

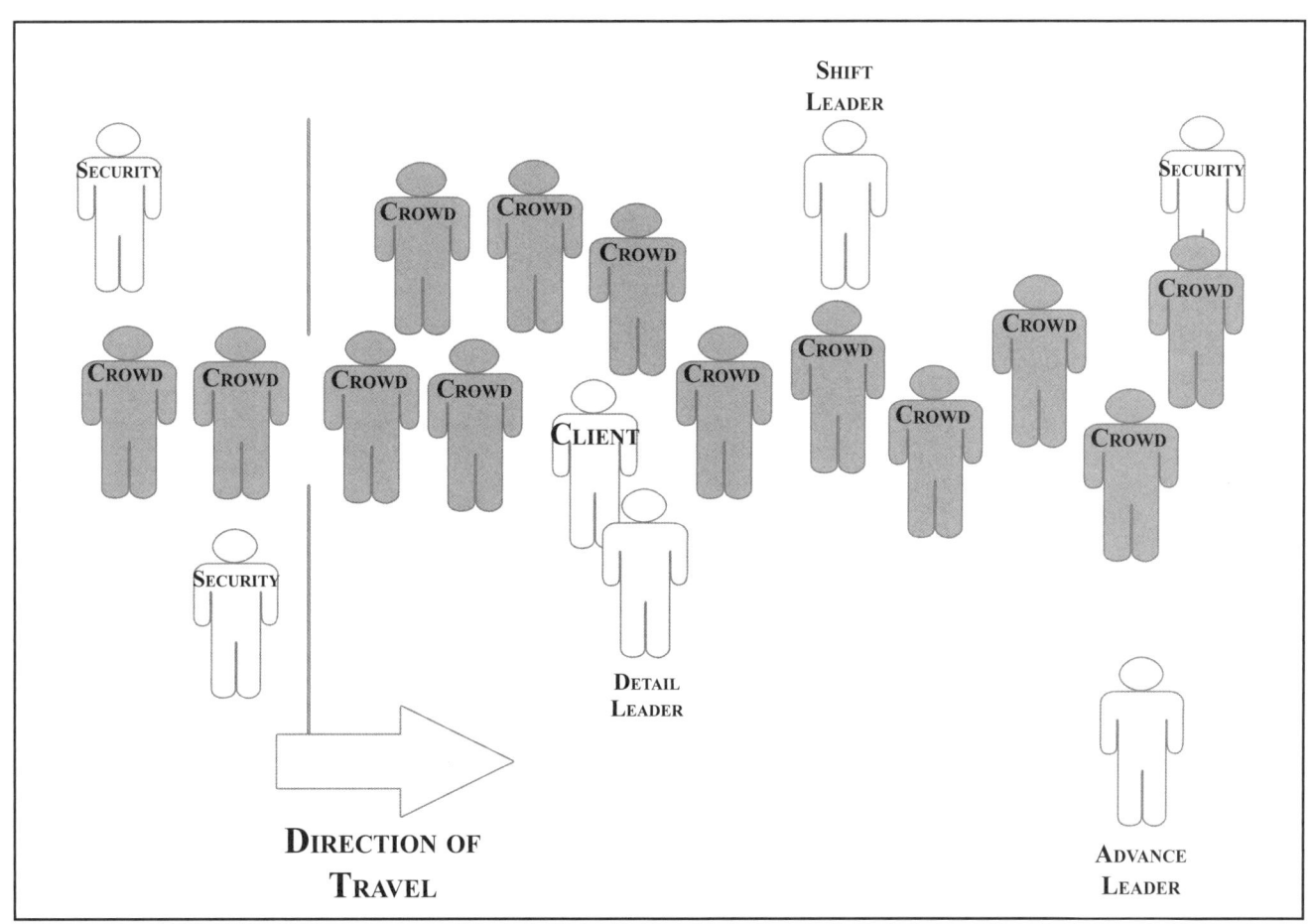

ENTERING ELEVATORS AND OTHER ENCLOSED SPACES

When taking your client into a facility where he will have to ride an elevator, make sure the advance is by the elevators waiting. When you exit the car, have them call all the elevators at the same time and hold them as you walk across the lobby using good 360 degree security. The advance will pick an elevator to use. When the detail arrives the client gets in first, then the detail leader, then security, shift leader and advance. The advance will be the first one out when the elevator gets to the floor since he will know where the client is going. There should be an advance detail member waiting on the floor where the client gets out to ensure security.   (See pages 97 - 99 for illustrations.)

One advance detail member and one security man will wait at the elevator on the first floor until the client is off the elevator and moving to his destination, then they will then take up positions in the lobby, waiting for the client and detail to come back down. Then the detail bringing him down will do the exact same thing. When they get to the floor for leaving the building, the two security men will stay behind, the advance and the security guys will be waiting to meet the elevator and pull security as the detail exits the elevator. The vehicle will remain in front of the building or the designated entrance until the client is on the floor for his meeting or other event. The car will pull back into position as the client is walking down the hallway heading towards the elevator. Remember the security sphere you decide to use is based upon your assessments, the in-house security, operations environment, publicity, etc. You might only need a bodyguard.

When you enter an enclosed space for transportation such as an elevator, bus, rail car or subway, you want to make sure you still provide 360 degree protection, just like when entering the elevator. Secure the doors, move the client in, then place the security around him, giving him layered protection. If you can keep everyone else off the platform your client is riding so much the better, but this is very unlikely, so you are basically building an in-depth wall of flesh between the client and all other persons using that platform for movement.

SECURITY SET-UP FOR APPROACHING AN ELEVATOR

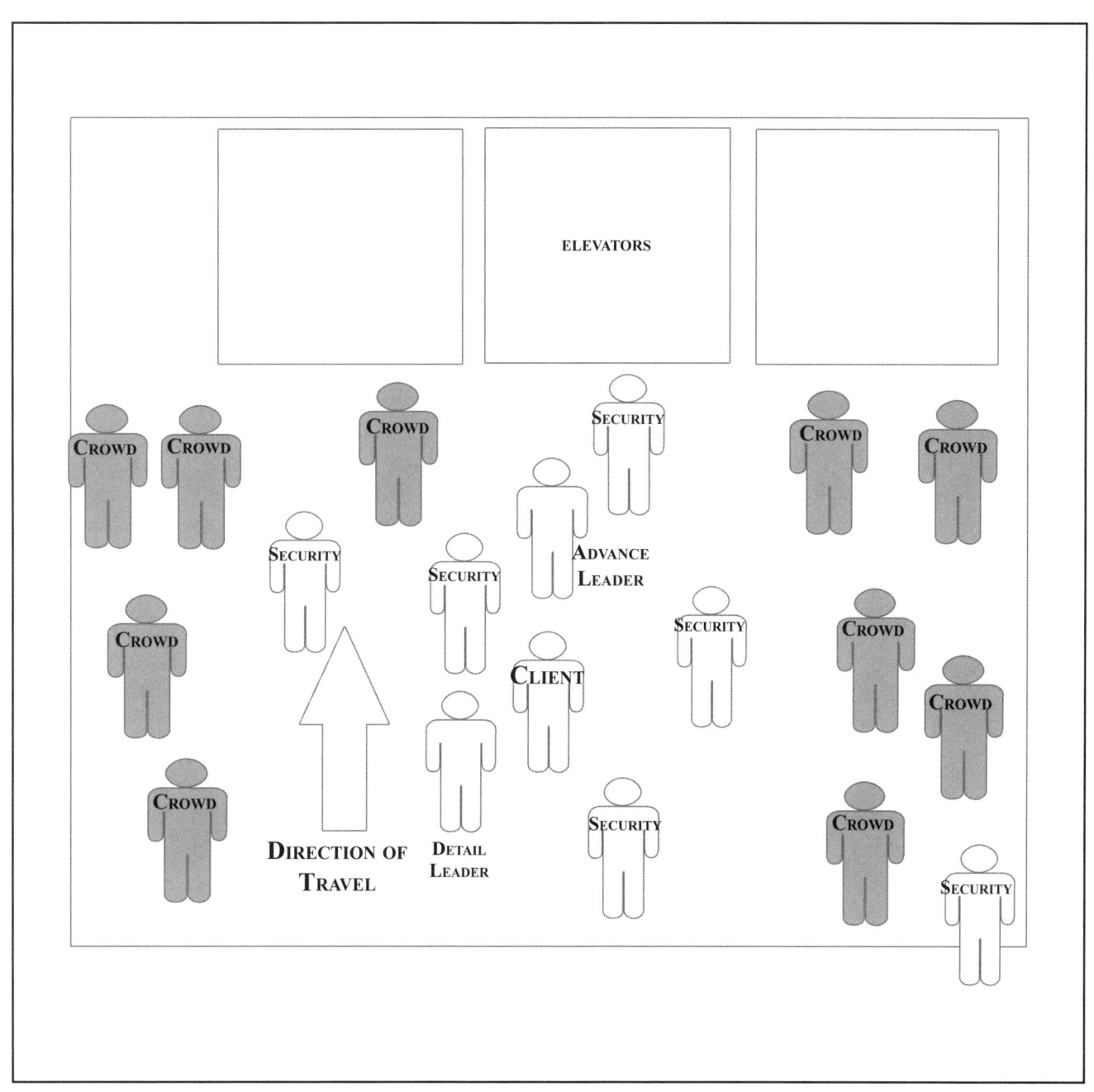

SECURITY SET-UP FOR ENTERING AN ELEVATOR

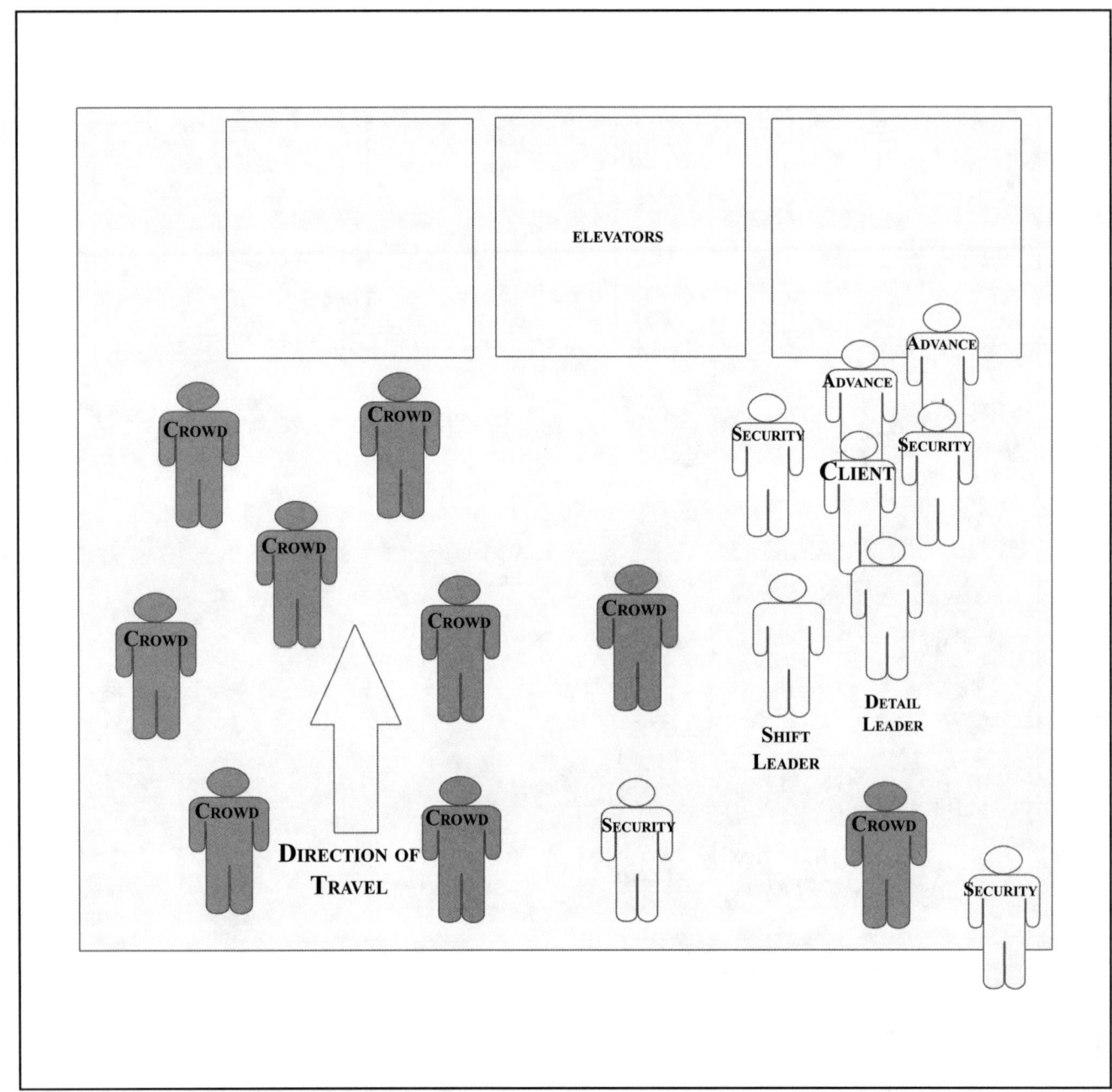

SECURITY SET-UP INSIDE AN ELEVATOR

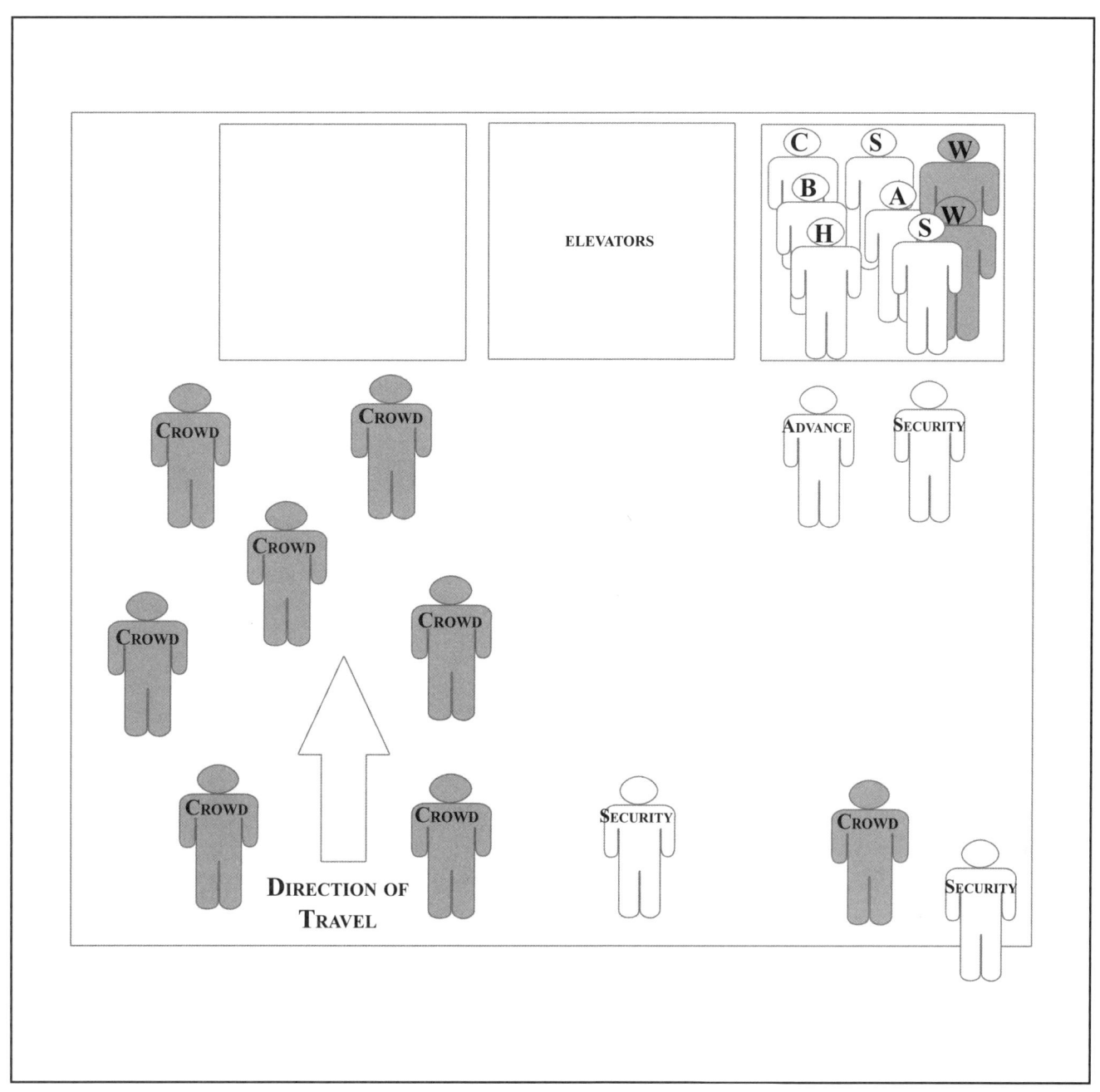

C = CLIENT   S = SECURITY   B = BODYGUARD   A = ADVANCE

H = SHIFT LEADER   W = CROWD MEMBER

NOTES

NOTES

CHAPTER 15
ROUTE SELECTION

The most dangerous time for the client and for you, his security, is when you are traveling to any site away from your secure location. Once you leave that secure location you no longer have any control of the environment and are moving through an area where potential threats are everywhere. The client more than likely will have to move to other locations/sites for meetings with subcontractors, reviewing work, checking work progress, etc. so you must know the needs of your client and plan for his schedule. After all, you wouldn't want him to show up late for opening a new facility with the press there, or be late for a meeting with a potential client because it could cost him work. It is very important for the security company/detail to understand the problems and potential incidents that can come from not knowing how to do proper route selection. If proper route selection is done, most problems and potential danger areas can be avoided, giving the threat no opportunity to strike at the principal or at you, his security.

The first thing you have to know is your client's requirements and schedule:

- Where is he going?

- What time does he have to be there?

- When does he have to leave either for another location or return to the office/headquarters?

- How many people are traveling?

- Are some people going on the movement staying at the destination?

- Are people leaving the location and traveling back with the client?

- Are there any other places he is visiting that day?

The client should give you all the information you need to complete the planning and the mission itself. That's what he hired you for, but he may not know what information you need, so you have to request it.

Always plan at least 3 routes to all primary locations of the client. Its very important that you are familiar with all the different places, locations and facilities that the client visits, as discussed in *Chapter 8: The Advance*.

When planning the routes of movement there are two types of routes for planning purposes: ones you are familiar with and have used before and ones that have not been used before (usually traveling to a new location). For known routes you should pull up the last after action report for when that route was used and study it. All drivers need to be familiar with the route. If it is one you travel on a semi regular basis (weekly) you probably won't need to send out a reconnaissance the day prior, but you should always check with military, local law enforcement, and other security companies to see if there have been any new developments, problems or incidents along the route.

A new route will be an unknown factor that needs to be researched and reconned before a movement is done on it. This should be done in low profile by all the drivers who are using the route, getting the timing and normal traffic patterns down.

There are three types of movements that you will need to plan for when preparing routes, including:

1. **REGULAR MOVEMENTS** are something done daily like a trip to an office or work area. For this type of movement you need to have as many different routes as possible, and some should link or overlap so you can switch routes during movement to help confuse any threat surveillance. You can consider using a security advance patrol or a counter attack team at choke points and possible attack sites to provide an extra layer of security. The regular movement is always the most dangerous type of movement since the threat will have picked up on the pattern, and will begin planning their attack from this known element of your movement profile. Vary routes. If you can't vary routes, vary times. If you can't vary times, vary departure points. If you can't vary departure points, vary arrival points. If you can't vary arrival points, vary times. The whole idea is to be as unpredictable as possible. This is necessary, because if the threat knows your destination and route and can predict a time, they can conduct an attack against you.

2. The **SHORT NOTICE MOVEMENT** is something the client has to do and was just notified of himself, so you will not have time for the regular route planning phase of an operation. Do a map recon and pick out the primary route and any contingencies routes and locate safe havens. You also need to inform the client if you have not been on this route before and you do not know its condition or roads. Tell him that you may get rerouted or lost causing him to be late so extra travel time can be figured in. It is always best to keep the client informed of what is going on. If you explain that a new route or that the movement on short notice may cause delays then he expects it. If you don't and he is late getting to his appointment, you look incompetent and he will start losing his faith in

your abilities.

3. A **SPECIAL MOVEMENT** is an event that is a one time movement to a special event like opening a school or a power plant, or meeting a VIP. It will probably be put out to the media so there will be coverage on it and you as security have to assume that everyone will know the time and place of the event and the people who will be there. You need to plan this out just like you would a regular movement and take all security measures you deem necessary. If you have enough notice you might even talk the client into going several days early to avoid the chance of an incident by being at the wrong place wrong time and avoiding the congestion of other VIPs, media and personnel who will be arriving for that event.

All decisions you make on routes and security precautions need to reflect the threat level, not only of the current working environment but all the areas you will be passing through and the final destination and return. This is when it is decided to go low profile or high profile for movement, to have a security advance patrol and a scout, etc.

ROUTE PLANNING PRINCIPLES

There are certain principals that should be followed when doing route selection and movement planning for the client. Like everything else in security everyone has their own way of doing this based upon their training and experience.

AVOID ROUTINE: Remember we do not want to be predictable or establish a routine that the threat can capitalize on but that is not always possible. This is dictated by the client since he might have meetings and events he has to attend and he does not have the ability to change them. This can also be controlled by the local military or law enforcement. In MOZUL, Iraq they would not let anyone travel on the roads before a certain time, and they sometimes enforced vehicle curfews when no vehicle could be on the road unless it was military at certain times. The threat, knowing these times, also knows when you have to travel.

MOVEMENT TIMES AND DATES: The times, dates, and destinations of planned movements should not be advertised to everyone. These items should be kept to a need to know basis. In some locations I have seen companies post movement schedules on the bulletin boards so people will know when and where they have to be, but unless you are in a secure location with restricted access this is not a good idea. Your times do not need to be exact and could fall within certain time windows. Uust because have to be at the departure point at 0900 doesn't mean you have to leave at 0900, as long as you arrive at the destination in time for the client to conduct his business.

PICK THE CORRECT VEHICLE: The vehicle used should depend upon the threat, environment, road conditions, weather, and number of personnel, but too many times companies just have one type or set of vehicles they use for every mission. This is not the safest way to operate and takes away the opportunity to change your security profile, such as doing a movement low profile if that would be your best option.

SAFEST ROUTE VS. SHORTEST ROUTE: You always want to travel the safest route possible to and from any locations; sometimes this will not be the shortest route to that destination. A lot of security companies get locked into using the shortest route because it is less time on the road so they have less time to be attacked, but once you become predictable and the threat knows where you are going to be and when, the amount of time on the road won't matter.

MAXIMUM SPEED POSSIBLE: This will depend on the type of environment and traffic you are traveling through, what type of profile you are presenting (high, low) and the client. If you are aggressive, driving as fast as possible, cutting people off and running traffic control points, people are going to notice you even if you are low profile. And driving fast sometimes upsets clients who are not comfortable with those types of things.

SECURITY ON THE ROUTE: Do you need extra security at choke points, danger areas, and other potential attack sites along your route? Is that extra security available?

COUNTER ATTACK TEAM: Do you need to employ a CAT, SAP or a quick reaction force while using the route to give you adequate security to meet the threat presented?

<u>*NOTES*</u>   <u>*NOTES*</u>

CHAPTER 16
THE PHASES OF ROUTE PLANNING

SELECTION OF THE ROUTE

An in-depth **MAP STUDY** should be done using the most up to date maps and satellite imagery available for that area. You are looking for wide roads, fast moving traffic with few reasons to slow down, good road surfaces, and routes that have the fewest number of danger areas or vulnerable points that can be exploited by the threat. Here is where you do your initial route break down, dividing the routes into stages, including how long each stage should be and how much time to drive it.

After you get your selected routes picked you need to send out a reconnaissance team to drive each route. They will drive each route at different times throughout the day and night to check that the stages' timing and distances are consistent, what speeds can be used, and which profile is best for the route. The route should be driven several different times at the same time of day it will be used with the client so that security can get a feel for the route, and it should be driven in the direction it is going to be used with the client. If it is only going to be used one way, that's how you drive it. If, however, it is going to be used both ways on the mission it needs to be driven both ways.

Certain information will be need to be collected by the route reconnaissance team including: how to gain entrance into the facility being visited, vehicle loading point, departure point, arrival point and vehicle unloading point, the selected routes, and alternate routes, high ground, danger areas and choke points. Areas that should be considered vulnerable points or danger areas include:

- Thick traffic, rush hours, or areas that have slow moving vehicles, lots of pedestrian traffic, road work or construction

- Traffic control points such as traffic lights, manned traffic intersections

- Areas with lots of parked vehicles lining the road

- Areas with no lighting

- Very quiet areas with no traffic either vehicle or pedestrian

- Traffic circles, intersections, alleys, on and off ramps at major roads

- Features that overlook the route or portions of the route, pedestrian bridges over the road, overpasses, hills, ridges, buildings, towers

- Anything that channels traffic, such as tunnels, bridges, roads with large ditches or culverts on the sides, narrow one-way streets

- Any type of traffic area that slows traffic or causes it to bottleneck

- Any areas that offer concealment to a threat such as thick trees, shrubs, high density pedestrian areas

- Steep roads, either going up or coming

down, that slow traffic, especially large trucks causing a loss of visibility of what's above until you crest that road

- Routes that have high banks on both sides or on one side of the road

- Roadways that narrow to one way traffic

- Areas that belong to criminal gangs, militia groups, or are controlled by terrorist organizations

- Routes that pass by large storage yards that contain combustible materials

- Weather conditions and how they affect the routes, such as flooding that slows traffic, roads turning into mud pits, etc

These all should have an influence on the routes you select and use when moving the client, and need to be checked out prior to any movement.

There are some additional things that need to be looked for when doing route selection these include:

ALL POTENTIAL SAFE HAVENS: These are places that could provide you with a safe location in case of emergency and have some type of protection. They include police stations, military posts, hospitals, banks, and jewelry stores.

MEDICAL AID: These are locations along the route that can give medical care in case of an emergency or unforeseen accident, including hospitals, medical clinics, and doctor's offices. You should have the name of a POC at each site, and contact information, and also a listing of what type of aid is available at each location.

COMMUNICATIONS CHECKS: You will need to conduct a communication check along the entire route, for your radios, cell phones, and satellite phones, as well as tracking devices and transponders on the vehicles. Make notes of all dead areas for each system, and what back up system will work in that area.

VEHICLE SERVICE SUPPORT LOCATIONS: For long trips you will need to know where you can refuel and check your vehicles. You also need to check what type of payment they take. Checking out garages, auto parts stores, and towing services would be prudent in case of an accident or vehicle failure requiring an emergency repair.

HELICOPTER LANDING AREAS/ZONES: If you have the ability to use air assets when necessary you should know where areas are along your route that can be used as HLZs, in case you need to use them to evacuate wounded or the client out of an area.

ORGANIZING THE MOVEMENT

SECURITY DETAIL OR TEAMS GOING ON THE MISSION:

Once the detail leader has made the movement plan and knows what his requirements are, taking into account the threat, the area, etc., he needs to start organizing the people who will be participating and supporting the mission and making liaison with any others who need to be kept in the know about the mission. This includes the clients so they know where they have to be, when they have to be there, what they have to wear and what they can bring and not bring. It also includes outside agencies like the military or law enforcement especially in high threat, high risk areas where they require any security details to send them a briefing on routes, times and locations so they know what armed people are moving in their areas. It also includes the security detail so they can prepare, and the advance, if any, so they can go do the site preparation work.

PEOPLE THAT WILL BE ASSISTING:

This will include any assisting force or organizations, including security at the locations you are going to, and refueling arrangements. It also includes anyone outside your normal security detail, such as extra personnel helping with site security at a location that doesn't have any or extra drivers when moving more than one or two clients. All these people need a heads up prior to movement time so they can prepare.

BRIEF THE SEQUENCE:

Get together with all necessary personnel, excluding the clients, to brief the sequence of events up to departure. This can include a reverse time plan. Make sure everyone knows when to be where and what the mission is and what equipment they will need, from mission start to mission complete (the return to station). You should also give out any travel details, maps, GPS way points, pictures and the latest intelligence for the area and routes to be traveled.

THE TRIP:

The detail leader needs to ensure that all timelines are met in sequence for the mission. This can mean making sure the advance leaves on schedule, the security advance patrol departs on time, the main body departs as scheduled and the counter assault team departs on schedule or the quick reaction force is ready on stand-by during the mission. The detail leader also needs to make sure the shift leader has all communications checks completed, weapons checks completed and that all vehicle tracking systems are operational. Then the mission is on and hopefully it will be a nice uneventful trip.

HOSTILE ENVIRONMENTS OR WAR ZONES

In this type of operating environment route selection becomes probably one of the most important things you do in protecting the client. The hazards, threats and other problems are increased enormously due to several factors that are not present in medium or low risk environments, including:

• **The movement of fighting factions and or warring parties throughout the area of operation** (Shifting lines of conflict). Hostile environments or war zones are inherently an unstable operating environment for several reasons: There can be local areas of control where people or parties control certain areas in a larger area that you need to pass through. Not only do you have to deal with those operating in that area where they may have check points or road blocks set up, but these areas can change very quickly during conflict so you are dealing with several different entities all with their own agenda or purpose. Sometimes they may just be criminals seizing the chance to extort some money. This can be very confusing to you and lead to uncertainty about what is a safe area or route and what is an unsafe area or route. This is why real time intelligence is so important, including using the security advance patrol to let you know what is taking place along the routes you have selected to travel.

NOTE: This includes not only hostiles but friendly forces. I have seen details denied access to routes, roads or cities by US or other coalition forces because they had operations taking place, or because the soldier in charge at that particular location hadn't been briefed on what to do with civilians traveling so he had no clue on what to do. Be flexible; Be prepared to deal with anyone or anything. DON'T GET PISSED!!

• It's possible to have both **marked and unmarked minefields** in an area of conflict. Many sides during a conflict will put down mine fields to channel movement in areas they control, or when they are in direct conflict, use hasty minefields to protect portions of the battlefield. In Bosnia this was done very frequently and restricted movement until safe routes could be found through these areas. Find out if, where and when this has happened in the area you will be operating in and moving through so you can make a better informed decision on your routes. If you need to send out the advance to scout the area looking for indicators of mine use, talk to locals. They usually know where minefields are located since they have to live there. This also includes any uses of cluster bombs or air deliverable mines. If these have been used in a way to restrict use of roads or areas you need to know. A good example of this is the cluster bombs the Israelis used in the 34 day conflict in Lebanon. The UN was still clearing them from fields and roads several months later. Or the use of them in the second Iraq war where some are still being found 4 years later.

• **Snipers or deliberate aimed fire** can be very effective in controlling movement in areas of conflict, especially in urban areas or built up regions. It is always best to have good working knowledge of the local area and up to date intelligence since you need to find out what areas are considered sniper alleys, and what type of people, and vehicles they are primarily targeting. In some cases they just try to hit military targets, in others law enforcement so this may be a low threat to you directly but there is also the chance that during your planned movement you could get caught in the cross fire between the sniper and his supporters and local military or law enforcement. You should also look at areas where the normal routine is harassing fire at any non-locals as they pass. This may be a low threat depending on your type of vehicle but a lucky shot can still hurt you and have a negative impact on your clients.

• **Vehicle Control Points/Vehicle Check Points (VCP)** can be used by anyone in an area of operations, including local militias, local law enforcement, local military, friendly military or coalition forces. They are wide spread in most hostile areas as a means to restrict movement, control the population and to slow and stop the flow of illegal weapons and munitions in the area of conflict. They will cause delays whether they are legitimate or not so the security advance patrol and good communications will be invaluable in locating and either bypassing or arranging for quick movement through VCPs. You should make sure everyone is at an increased level of awareness as you approach these and that everyone knows how to recognize a legitimate check point. Vehicle Control Points have been used to steal vehicles, kidnap people, rape women and to execute people who are not in line with the people who control that checkpoint. Before leaving your start point, KNOW where all legitimate check points are, call in any others and proceed with extreme caution.

• **Civilian Pedestrians in the area, including PROTESTORS**. People fighting for rights or extremists always seek publicity to get their point of view heard on television or in the media. In a hostile environment or war zone these kind of publicity events are amplified in their importance because of the environment they are taking place in.

In some areas or countries at certain times of the year there will be **RELIGIOUS PILGRIMS** making necessary journeys for their beliefs. They usually walk but have many support sites and vehicles along to routes they use to help them. They block roads, they slow traffic, they may throw things at you. Know the routes they are using and avoid these areas until the time of their movement is over.

In some countries, including the US, **SPORTING EVENTS CROWDS** are very enthusiastic. This is true even in hostile areas and war zones, and can lead to gun fire, fireworks, large crowds in the streets and other shows of support. Anyone in Iraq during the last Olympics knows that displays of support can take many forms. Watching the tracer fire as the Iraqi soccer team was doing well was an interesting sight. Know what is going on with these types of things and know WHEN. Plan your movement and routes away from any areas where there may be some disturbances.

There will be some **HOLIDAYS** that are celebrated by events that take place in the community, including parades, shows, or food fights taking place. It is a good idea to use your local knowledge and intelligence for finding out when holidays are and how people celebrate them so you can plan your missions around the dates and if you cannot change the dates at least know what areas to avoid.

Even during normal times **FUNERALS** are very emotional events, but it is taken to the extreme in hostile areas or war zones. You should routinely avoid graveyards or morgues during your trips. You need to rely on your security advance patrol to let you know if there is a funeral taking place along your primary route, and if there is you might want to go to an alternate since you don't know how the person died or who killed him. You might become a target of hatred just by being there.

NOTE: There has been an ambush tactic used in the past where large crowds were deployed very quickly in an urban environment to slow and or stop traffic so that people could be pulled from vehicles. Know if there is a scheduled event and be suspicious of anything not planned for or known.

• **Sustained areas of fire** are a very real threat in a hostile working environment or war zone. Fighting factions could be having a fire fight during the time of your movement, they could be putting on a fire power demonstration for the media or supporters, or they could just be doing harassing fire on anyone who is not a member of their particular faction to prove politically that they control their areas. Another problem that could be added to the mix is the fact that there could be outside peacekeeping forces in the area and they will be as confused as you are, so you have to deal with them also. Up to date intelligence, real time intelligence and good communications are important when doing movements in areas like this. You must have other routes scouted out and preplanned in case you cannot use your primary. Being aware of the shifting political climate it is also important. I used to have my translator bring in all the local papers and translate them for us so we could get a feel for what the different factions are saying. He would also translate all talk done over local loud speakers, and local radio shows. You need to know what the locals are thinking and what they believe.

You need to minimize the risk of possible attack against your client whether it is intentionally planned or he is just a target of opportunity. You do this by having security in place with 360 degree protection and a good understanding of the threat. Movement is the most dangerous time for you and your client so it is vitally important that you chose those routes and times that present the least overall danger to the client, and that you pick the proper equipment necessary for the mission and the proper security profile to ensure your client never becomes a target or victim of an attack.

NOTES

NOTES

CHAPTER 17
MOVING THE CLIENT

IDENTIFY THE ROUTES

The security detail will need to identify primary and alternate routes to all daily primary locations, and to all areas to be visited by the client. The best routes to choose are ones that will have ease of movement during the time of movement, where there are few or no stops such as construction or other traffic control points, non-winding roads, and few danger areas and choke points. Once the primary and alternate routes are identified, highlight both routes on a map or map overlay for the entire detail to study and become familiar with.

Next you will need to identify all potential safe havens along or near those routes throughout the entire length of all routes. What exactly is a safe haven can and will depend on your location, and the type of environment you are operating in, but some suggested safe havens are Police and Fire Departments, hospitals, military check points, military bases and banks. Once these potential safe havens have been identified coordination with them needs to be done. Visit each site and see if it will function as a safe haven. Find out who the person in charge is, get contact numbers and a point of contact you can speak with if there is a problem, and find out what it takes to gain access during an incident. All verified safe havens as well as potential safe havens should be marked on a map or map overlay; everyone in the detail should become familiar with them and their locations.

While you are doing your map recon and driving the route, you should be identifying choke points and danger areas. Tunnels, construction sites, narrow winding streets, traffic lights with long wait times, bridges, culverts, blind curves on hilly or mountainous terrain, on and off ramps, overpasses and congested traffic areas are some of the danger areas you need to look for. Areas you MUST pass through to get to a site location on a frequent basis are choke points. Both choke points and danger areas should be marked on the map for the security detail to study. This is important as it will tell them where and when they need to heighten their awareness and observation for possible attack.

Next the detail needs to drive each route during the different times of the day they may be traveling, including on weekends. If possible they need to drive them in inclement to identify their actual travel times to locations, and on the ground distances to and from each site as well as the distance to each safe haven. Once this information is collected it needs to be marked on a map or map overlay for all of the routes so it can be used for mission planning.

You will also need to identify the hospitals along your route, not only as possible safe havens but their condition specific capabilities, including time and distances to those hospital or clinic sites. It is important to find out the condition specific capabilities of each location because while some hospitals may be acute care centers and can handle any condition or trauma some may not be able to treat all conditions, including one that affects a client or clients. Most, however, should have emergency rooms and they can treat minor injuries and stabilize life-threatening injuries. Some hospitals in the area may be full trauma level I (preferred) hospitals that can treat all conditions. When selecting a hospital during an incident while participating on a mission, the team medic will need to consider the type and severity of the injury and what additional medical attention is needed. The locations and type of care available at the different hospitals or clinics and the times and distances to the each site should be marked on a map or map overlay for the detail to study.

The drivers, and if possible, the whole detail should rehearse the driving of the routes by actually going out and doing it prior to the mission. I would suggest doing it in low profile vehicles so as not to alert any threat surveillance what the possible route for your next mission will be. Always rehearse driving the routes and, if time permits, the absolute best time to rehearse driving the routes is the same time of day that the route would be driven on the mission day. This can be very important because in some high density urban areas and other major metropolitan cities, two-way streets change to one way for rush hour traffic. You need to check if this applies to your selected routes.

CHAPTER 18
MOTORCADE OPERATIONS

TERMS AND DEFINITIONS

• **MOTORCADE/CONVOY**. A formally organized group of motor vehicles traveling along a specified route in a controlled manner.

• **PILOT VEHICLE/SCOUT**. When the resources and manpower are available, a vehicle will be used for scouting or conducting a recon on the route for the motorcade. Also called a Security Advance Patrol (SAP), which can act as the CAT vehicle when necessary, this vehicle is usually a low profile vehicle making it easier to blend in and not draw attention to itself and the detail following behind.

• **LEAD VEHICLE**. Takes the lead position in the movement directly in front of the main body vehicle and is responsible for front and flank security.

• **MAIN BODY**. The vehicle in which the client or clients rides. It is driven by a person who is thoroughly familiar with the entire geographical area and is trained in defensive driving techniques. There may be more than one in a convoy or motorcade so you could have Main Body 1 and Main Body 2.

• **FOLLOW CAR/TRAIL VEHICLE/GUNSHIP**. A security vehicle driven directly behind the main body. This vehicle is responsible for protecting the main body from the rear and the flanks.

• **CAT VEHICLE**. A security vehicle that is driven about 100 meters behind the follow car, or could be in front of the lead vehicle depending on the needs of the movement. This vehicle is responsible for protecting the main body from the flank at halts. It is also responsible for quickly interdicting a threat by placing itself between the threat and the clients in the main body vehicles. It is controlled by the shift leader who will tell it when to engage. Usually a low profile vehicle so it blends in.

• **ADVANCE.** This element coordinates for and recons planned and unplanned stops. The advance leader should be one of the most PSD experienced members of the detail.

MAIN BODY DRIVERS:

• Ready to move at a moment's notice.

• Reacts to threats by taking directions from the detail leader as necessary.

• Thorough knowledge of all routes, primary and alternate, day and night, and location of safe havens, hospitals and police stations.

• Ensures the vehicle is clean inside and out, and in good mechanical condition (follows vehicle checklist).

• Responsible for the security of the main body, during stops.

LEAD VEHICLE DRIVER:

- Same requirements as main body driver.

- Protects main body from front.

- Acts as a spare main body

- Thorough knowledge of all routes, primary and alternate, day and night, and location of safe havens, hospitals and police stations.

- Ensures map is available.

FOLLOW VEHICLE DRIVER:

- Same vehicle requirements as "main body" and "lead".

- Serves as communications link to security operations center.

- Responsible for rear and flank security.

- Operates tactical radio during shift leader's absence from vehicle.

SUPPORT PERSONNEL:

- Advises driver as to traffic conditions around motorcade.

- Familiar with employment of all weapon systems and their assigned sectors of fire, both mounted and dismounted.

- Takes instructions from shift leader

RIGHT REAR. Security person seated in the right rear of follow car.

CENTER REAR. Security person seated in center of rear seat of follow car.

CURBSIDE. Location of client's arrival.

SITE SURVEY. Investigation and resultant plans of security for given location.

SWEEP. Search of an area to determine all possible threats.

SECURITY OPERATIONS CENTER. Communications center that ensures all information is communicated to proper individuals in protective detail.

SECURITY PERIMETER. Unit that involves personnel, alarms, barricades and other devices to provide a secure protected site.

SECURITY POST. Area of responsibility established to form part the security network.

You should always make sure when you plan your mission, especially on long trips, that you have enough space in the vehicles to carry everyone if one vehicle is out of service for some reason. This means if you are carrying 6 clients, and you only want 3 clients per vehicle that you need a main body spare in case one of the main bodies goes down. If one of the security vehicles goes down you need the space to cross load them and any vital equipment into the other vehicles. NOTE: I have seen convoys go out with the bare minimum and one vehicle broke down, then the entire convoy including the main body had to wait until another vehicle could be brought out to tow the disabled vehicle back and they had to bring a spare so the mission could continue on, leaving everyone exposed in order to cut back on costs.

VEHICLE EMBUS AND DEBUS PROCEDURES

Some of the most dangerous times of any movement are picking up or dropping off the client, when the client is getting into the vehicle for departure (EMBUS), and when the client is getting out of the vehicle (DEBUS). At these times the vehicle is stopped and the integrity of the armor is broken because doors are open. The client and his detail will now move from the protected environment of an armored vehicle to movement by foot to enter the destination facility or building. Most entry sites are to the front of a building so now the threat has a location to plan for and once they know this location, they have a way to attack the client. You should never set patterns. If you have to visit the same site on a regular basis use every door in the building to debus the client, not just the front door. You can also use underground parking if it is available or the delivery area. A thorough knowledge of the site and proper planning will allow you to keep from setting a pattern that the threat can use to their advantage.

There are several ways to get into and out of the vehicle. The first is the normal way (orthodox) where the client exits the vehicle through the door on the same side as the entrance to the facility. The route is clear and he just has to step from his door and walk into the front door.

The second way is the special circumstance, or unorthodox way, when the door the client is using to exit the vehicle is in the rear of the vehicle or on the opposite side of the vehicle from the facility entrance. This means the client will have to walk in the street or around the vehicle to gain the sidewalk and enter the facility. This increases the distance the client has to move to get off the street and into a protected site and increases the time an attack can take place. It also removes the armored vehicle he is exiting as a barrier of protection until he is off the street and around it, heading towards the entrance.

EMBUS (GETTING IN) CONSIDERATIONS FOR SECURITY

Make sure the detail personnel and vehicles are in place before the client is scheduled to arrive. It looks unprofessional for a client to show up and see the detail standing around smoking or urinating on tires. It also increases the threat to the client since it will take longer for the mission to depart if he has to wait for everyone to load up or for the vehicles to show up. When outside a secure pick up area the main body needs to be parked as close to the door as possible. Less distance between the door and the vehicle means less time for the threat to attack.

The detail vehicles and personnel should be ready and the main body driver should have the car started and in gear, so when the client gets in, the detail leader can close his door and enter his vehicle then the security detail can leave their positions of observation and enter their vehicles and the convoy can start out. This needs to take place very quickly. Any briefings to the client can take place on the movement and buckling up can be done once the vehicle is underway to lessen the amount of time the client and the convoy are stationary targets.

The shift leader needs to inform the detail when the client is exiting the facility then the detail will need to keep eyes out, scanning their area of observation until the client gets in the main body and the detail leader tells them to mount up. This is difficult as it is human nature to look around at voices, noise, or just to see what the hell is taking so long. Any baggage or packages accompanying the client should already be loaded before the client exits the secure site. The client gets in the vehicle, the door is shut by the detail leader or bodyguard, and the driver locks the doors as soon as the client's door is closed. The detail leader or BG should not be carrying anything for the client. His hands should be free, and he should not hold anything for the client when he is entering the vehicle. Also be aware that when you have female clients, keep your eyes off the cleavage and the legs when they are entering the vehicle.

DEBUS (EXITING) CONSIDERATIONS FOR SECURITY

The detail leader needs to make sure that the main body stops as close as possible to the facility with the client's exit door facing the facility door so the client has a shorter distance to walk to the entry point. This lessens his time as a possible target. The advance should have security set up for the arrival. The convoy security detail needs to get out and help when the convoy arrives. The client should wait until security is out and controlling the area before exiting the vehicle.

The convoy vehicle drivers stay in their vehicles with the engine running and the vehicle in drive with their foot on the parking brake. The wheels should be pointing in the direction of the escape route not straight ahead or at another convoy vehicle, so that if an attack occurs during the debus procedure, the client can get in the vehicle and everyone can evacuate the area. There should already be preplanned escape routes identified by the advance and briefed to the drivers before the mission start.

The detail leader or BG will open the client's door after security is in place: once again the detail leader does not carry any bags for the client or hold things for them while they exit. Any baggage or packages going in will be taken inside the facility after the client is inside, and once again for female clients, do not look at cleavage or legs when they are exiting the vehicle, and watch your hands if you have to help them in and out of their body armor.

When the security detail exits their vehicles to set up a secure perimeter in preparation for the client to embus or debus they will need to place themselves where they can see as much as possible in their assigned areas. They need to identify fields of fire, cover and a route to that cover in case of an attack. The security detail may have to stop pedestrian traffic while the client moves from the vehicle to the facility entrance.

The advance, if there was one, should already have security deployed when the convoy arrives. If so, they will direct the convoy security detail where they need to stand to increase the sphere of protection for the client. If, as the detail leader or bodyguard, you are going to open the door for the client, do not open it so the door is between you and the client. The door is cover, so you need to stand so that the door is on one side and you are providing cover on the other, protecting the client from two sides until he starts moving. If there is the threat of a sniper attack, you can have a black umbrella ready to shield the client from observation and aimed fire as he moves into the facility.

After embus or debus is completed the vehicles move to their staging area. The vehicles should not be left unattended at any time and this is the time that the driver should refuel and take whatever actions he needs to get the vehicle ready for the next movement. Make sure these actions are staggered so all the vehicles are not gone at the same time. The emergency evacuation vehicles should be staged at each designated exit for the evacuation plan as soon as the clients are in the building.

REACTING TO AN ATTACK WHILE MOVING

There are only so many things that can be done to get out of an ambush or attack designed for vehicles.

While the client is being moved by vehicle there are three options for the security detail who are protecting the client:

1. Remove the threat, by neutralizing it in some way, kill it, scare it off, etc.

2. Protect the client by placing security between him and the threat. This can be done with the security vehicles, taking the client into a secure facility, using smoke or flash bangs.

3. Move the client away from the threat; this is the simplest - just drive away - and should be the done as soon as a threat is detected.

Any counter ambush drills you conduct with vehicles have certain basic things that have to be accomplished:

1. Protect the client at all costs. It is your job to drive away, not stay and engage the threat, or to try to recover other security personnel.

2. Remove the client from the danger as quickly and safely as possible. It does no good to get out of an ambush only to die in a car wreck.

3. Take violent, aggressive, fast action to counter the unforeseen threat. This will surprise the attackers and help you regain some control of the incident.

4. Do not just react, THINK, analyze the situation. Why was the attack here? Is it a decoy? Are the attackers trying to channel you into a better kill zone?

SCENARIO: When attackers have the way partially blocked you have to drive through it:

ONE VEHICLE OPS

• The detail leader will make everyone get down below the level of the windows where the vehicle armor is strongest and the attackers cannot aim their shots.

• If possible, the detail leader will return fire if he does not have to break the armor integrity of the vehicle.

• The driver will ram the barricade or vehicles blocking the route of movement and accelerate away.

• Inform the security operations center as soon as possible.

• Re-assess the situation, keep the clients down and move to a safe haven.

TWO OR MORE VEHICLE OPS

The main body and detail leader or BG perform the same actions as for a one vehicle op. The security vehicle will conduct the initial ramming of the threat barricade or vehicles, then the main body will follow, accelerating past the security vehicle while the security vehicle places itself between the threat and the main body, returning fire if appropriate and civilian damage can be kept to a minimum. As soon as the main body is away, the security vehicle will attempt to catch up.

SCENARIO: When the attackers have the route to the front completely blocked:

ONE VEHICLE OPS

• The detail leader will make everyone get down below the level of the windows where the vehicle armor is strongest and the attackers cannot aim their shots.

• The driver will bring the vehicle to an immediate halt as safely and as quickly as possible.

• If possible, the detail leader will return fire if he does not have to break the armor integrity of the vehicle.

• The driver will reverse out (no forward or reverse 180's) as safely and as fast as possible.

• Inform the security operations center as quickly as possible.

• Re-assess the situation, keep the clients down and move to a safe haven.

TWO OR MORE VEHICLE OPS

• The main body will do the same thing as for a one vehicle op.

• The lead security vehicle will move between the threat and the main body, returning fire if appropriate as the main body reverses out.

• Once the main body has reversed out and is moving away the security vehicle will reverse out and follow.

If there are 3 or more vehicles everything is done the same by the main body and the lead. The rear security will reverse out first, followed by the main body, then the lead security.

SCENARIO: When the attackers have the front completely blocked off and you cannot ram through and the rear is blocked off also:

ONE VEHICLE OPS

• The detail leader will make everyone get down below the level of the windows where the vehicle armor is strongest and the attackers cannot aim their shots.

• The driver will bring the vehicle to an immediate halt as safely as possible.

• If possible the detail leader will return fire if he does not have to break the armor integrity of the vehicle.

• The driver will reverse out (no forward or reverse 180's) as safely and as quickly as possible.

• The detail leader needs to determine where the main threat or concentration of fire is coming from and have the driver position the vehicle as far away from the threat as possible and at an angle.

• All personnel will exit the vehicle on the side protected from threat fire. Keep the client on the floor of the main body as long as possible (best protection).

• To regain some control of the situation and gain some breathing space and respect from the attackers, the security personnel need to do a mad minute at the threat.

• Use smoke and flash bangs to help screen aimed fire at you and help identity your location to the SAP or CAT.

• Find a safe haven.

• Inform ops as soon as possible about situation and location.

• Evacuate the client when possible.

TWO OR MORE VEHICLE OPS

• Both vehicles will move as far away from the main concentration of fire as possible and stop at an angle with the end of the vehicles touching, forming a rough V shape towards the threat.

• Everyone exits the vehicles on the protected side. Leave the clients on the floor of the main body for the most protection.

• To regain some control of the situation and gain some breathing space and respect from the attackers all the security personnel need to do a mad minute at the threat.

• Use smoke and flash bangs, to help screen aimed fire at you and help identity your location to the SAP or CAT.

• Find a safe haven to move the client to on foot using aggressive fire power.

SCENARIO: The main body is immobilized or damaged by threat action in some way and you need to use another vehicle to evacuate the client so you move the client to an undamaged vehicle to evacuate him from the area.

• The detail leader or BG makes the clients get down below the level of the windows in the main body vehicle.

• When the security vehicle is ready they will radio the main body. At this time the BG will exit the vehicle with the client, using the vehicle for cover and move away from the vehicle allowing enough room for the security vehicle to pass between him, the client and the main body. (This will give two layers of protection for the client and performing the exchange this way will keep the integrity of the armor facing the threat intact.)

• The security driver will pull along side the main body, placing the main body between it and the threat firing positions.

• If necessary, security people will exit the security vehicle to give the clients room to enter and the driver of the security vehicle and shift leader will stay in the vehicle.

• The clients are moved to the security vehicle with the detail leader or bodyguard.

• The security vehicle driver will then drive out of the kill zone to the nearest safe haven.

• The security left behind will stay and engage as long as possible before moving on foot out of the area keeping the security operations center informed so they can get picked up by the CAT or quick reaction force.

CONSIDERATIONS

• The occupants of the main body could be stunned or unconscious, so someone should be designated to extract the client.

• Make sure you are not too close to the main body when you pull along side. You want the doors to be able to open.

• Who will ride with the clients in the security vehicle when it pulls away? It is better to designate this before an incident so everyone knows who will be performing this function.

THINGS TO REMEMBER

• The vehicle armor does not make you invincible, it protects you for less than 2 minutes from sustained small arms fire.

• Your vehicles will draw attention and fire.

• It is always better to make a decision and carry out a bad plan than to sit and consider the perfect plan in a kill zone.

• Be flexible in everything.

• A vehicle blocking you to the rear may be an innocent.

• When moving the client from the vehicle place him behind a vehicle strong point such as the wheels or engine block for protection. If the vehicle is immobile anyway you might want to consider shooting the tires out so the rims rest on the road, giving more protection and lowering the frame of the vehicle closer to the road.

• The shift leader or senior security guy should decide when to use smoke or flash bangs.

• Be violent, be fast, and be aggressive in your response but be CONTROLLED.

• Doing the wrong thing to protect the clients is better then doing nothing at all.

WHAT COULD GO WRONG

• Exiting vehicles on the side that is taking fire.

• Not closing doors after exiting.

• Poor weapons handling.

• Poor communication, everyone trying to talk at once.

• Staying in the kill zone too long.

• Not using smoke.

• Returning to a downed vehicle after moving away from it to retrieve bags, ammo, radios, gear, etc. If you didn't take it with you when you got out, it must not be important so SCREW IT, leave it.

• Lack of flexibility. If an SOP isn't working, don't keep doing it, try something else.

• Being afraid to make a mistake (no initiative). You're under fire in a kill zone DO SOMETHING!

If nothing else, just remember to get out of the attack area away from the threat. Your three major options are probably the easiest too remember:

1. **Drive straight through the ambush** which is probably the best first option and the one that will get you out of the ambush the fastest. Just remember to get everyone inside the main body vehicle down below the level of the glass and floor it, do not hesitate. A fast aggressive response by the drivers at the beginning of the ambush may catch the attackers off guard, but if it is a well planned ambush this may not be feasible.

2. **Ram through a vehicle obstacle**. Once again a quick decision that needs to be made by the driver as soon as the ambush is recognized. Most vehicles can withstand ramming another vehicle if it is done correctly, and it disrupts the attacker's plans.

3. **Back Up.** The last major option is to reverse out of the ambush area if you cannot or decide not to drive through. The main body or vehicle carrying the clients should be the first out while the security provides over watch and cover fire until the main body is out of the area, or the clients are evacuated into another vehicle and that is out of the area.

CHOOSING THE RIGHT VEHICLE FOR THE MISSION

You need to pick the right vehicle for the jobs you are going to do and the missions. Many security companies have a one vehicle fits all policy, which is cheaper and better for them but not necessarily so for the client. There are several things you need to consider when determining what type of main body vehicle you will need to transport the client in and what type of vehicle to use for the security elements.

• What is **financially available**? Sadly this probably the largest factor in determining the type of vehicles that will be used.

• The **manpower** you have available for the mission's types of convoys.

• The **assessed threat level** which should be done for each mission to determine the type of vehicle and security profile you will use. Once again usually security companies have a one size fits all for every mission.

• Are the **drivers** trained and do they have experience with the types of vehicles that are being used? For some types of vehicles you need to have experience with the extra weight and engine power because these will change the handling characteristics of vehicles.

Some of the characteristics the vehicle should have are:

• At least **4 doors.**

• A **large engine**, V8 or better, that has some power behind it (if you have a frame up modified vehicle with armor the engine should have been upgraded also to handle the extra weight).

• **Spare tires** need to be easily available, not only while on the vehicle but replacements for them after each incident, so you need to have a lot of extra tires on hand or they need to be readily available in the local economy.

• All tires should be **run flats**. While run flat tires are not necessary for a vehicle they do provide better control for the driver during an incident, because the driver will have control of the vehicle for longer.

• A **central locking system** for the door locks to ensure all doors are locked at the necessary times.

• A **self sealing fuel tank** to prevent the loss of fuel, and a possible fire, during an incident.

• High profile/high visibility vehicles need to have **reinforced bumpers** that have the ability to take impacts and push other vehicles.

• For low visibility vehicles if **reinforced bumpers** can be added without totally changing the look of the vehicle then do it. If it makes the vehicle stand out in any way, don't.

• **Larger side mirrors**, with smaller mirrors placed on them to increase the outside view, and larger rear view mirrors should be placed in the vehicles.

• A **mirror for the passenger side** should be placed on the visor. Remember "the more you can see the safer you will be."

• **Windows should be tinted** in all the vehicles, even the front. This will restrict the ability of others to see into the vehicle and should not restrict anyone's ability to see out. In some places tint is unusual but window curtains in vehicles are not, so you may have window curtains in your vehicles especially in low profile vehicles.

• **Windows should be bullet proof** in all vehicles and have the ability to roll down.

• **Air conditioning** that works will keep you from having to have your windows down during transport which is not good since having the windows down takes a way a layer of protection.

• On high profile vehicles you should have **extra protection on the roof,** such as Kevlar blankets laid out on top and tied down. Some also carry several spare tires up there to help slow down or deflect any rounds fired from the tops of buildings or on highway overpasses. The carrying of tires on top of vehicles is an on going debate. I believe it is a good idea as long as one spare is located inside the protection of the armor. The tires are carried up there so you can do a faster tire change if necessary. If you are taking rounds from an elevated position you are not going to stop and change your tire until you are in a covered or protected area anyway so that extra layer may save a life.

ARMORING VEHICLES

This is another debated topic. Some people believe that the only factory armored vehicles should be the main body vehicles or just the ones carrying clients. Some, like me, believe all the vehicles in the detail should be factory armored, while others believe vehicles carrying the security detail don't need armor or can be self armored or modified away from the factory. When a vehicle is armored by the factory or a company that specializes in doing it, then the engine, transmission, air conditioning, shocks, tires, and brakes are all done to handle the extra weight of the vehicle so it will not lose any performance value. The door hinges also have to be modified so the extra weight on them does not cause the doors to sag or the hinges to warp over a period of time. Also many times the factory can put in bullet resistant glass so that the windows still work to some extent. This is necessary especially for the driver if he has to pass information or talk to people at check points so he doesn't have to exit the vehicle or break the armored integrity of the vehicle too much Bullet resistant windows are also important because they are a way out of a vehicle that has been damaged and the doors jammed. They can be busted by sledge hammers and people extracted.

Locally armored or vehicles armored by others after purchase have several problems, especially for the SAFETY of the occupants. First they usually add plate steel to a purchased vehicle around the doors, drivers, and between compartments. They cut out portions of it so that the security detail members can roll down the window to see out of and to shoot out of. These openings are usually way too small for a man to pass through, especially a man wearing kit. The added weight will cause problems with the transmission, engine and cooling system because it will be outside the factory specifications for that weight. The doors will start sagging and dragging down on the hinges and the door frame because of the added weight, and door locks and door handles are usually blocked by the added protection and not easily reached if it becomes necessary.

ARMOR: STEEL PLATES ADDED TO THE DOORS

This is one way that companies modify their vehicles for security. They add steel plate doors on to the vehicle for that extra layer of protection and leave standard glass in.

This provides no protection from the front of the vehicle and a person could not crawl or be pulled out from one of the side door windows because with the protection in place there is not enough room.

With this configuration a person could crawl over the seats and exit through the windshield if necessary. It would have been very easy to put the top part of this plate on hinges so it could be lowered if needed.

ARMOR: ADDITIONAL STEEL PLATES AND KEVLAR

This is another way to modify vehicles and while this one looks better it is actually worse.

Right behind the driver's seat is another set of steel plates separating the front from the rear, and on the sides of the gunner's box. When the back is closed off for rear protection the steel goes up to the roof.

There is no way to get someone out if the doors are jammed or inoperable. People in the front can beat out the windshield but no one can get out from the rear or side doors and windows.

Notice the Kevlar blanket on the roof of this vehicle for the extra protection. Once again if they had used bullet resistant glass or put portions of the steel plate on hinges so it could swing down out of the way it would be much better for those using this vehicle.

HIGH PROFILE VEHICLES

Many companies like to employ what I consider to be very high profile vehicles that look military. They are used by military or tactical law enforcement in other countries and now are being purchased as a way to transport clients in high risk areas or war zones or used as security vehicles. Like every vehicle out there, you have pros and cons. I am not going into the pros of using these types of vehicles but I am going to discuss some of the reasons I believe they are bad not only for clients but for security use.

1. A B6 Level armored vehicle whether it is a sedan, SUV or one of these military type vehicles all meet the exact same specifications, which is they can withstand 7.62x51 type rounds fired from a distance of 10 meters. Many security companies try to say that these high profile vehicles can take more but this is not true unless it has a higher rating than B6.

2. The frames on these vehicles are set much higher than the frame of standard vehicles, including SUVs. This means that unless all vehicles in the movement are the same, a standard vehicle could not be pushed from a kill zone by one of the high profile vehicles or push one since the difference in bumper height is too great. Also towing will be very difficult and the weight difference might also cause problems between the high profile vehicle and the SUV.

3. Evacuating people while under fire usually means exiting the vehicle on the opposite side of the vehicle from where the fire is coming, keeping the damaged vehicle between you and incoming rounds. In these high profile vehicles there is usually one single entrance in the back of the vehicle which means clients and security personnel would have to exit under fire regardless of the direction of attack.

4. These vehicles usually come with bench type seats placing people with their backs along the windows of the vehicle, which restricts those people from being an extra set of eyes during the movement to help look for trouble or suspicious activity. It also allows for better targeting of those inside the vehicle since there is a much clearer view, and the widest part of their body is exposed to weapons fire or shrapnel from IEDs, versus the smallest when sitting in profile.

5. Most of the wheels for these vehicles are HUGE, because of the extra weight and height of the vehicle. This makes it very difficult to change a tire quickly, because of their weight and size but also because they have 20 or more lug nuts. Also ground clearance on these vehicles makes them very difficult to jack up.

6. The height of the step from the ground for getting in and out of these vehicles is high. Not every client is in the best of shape or tall, and those will have a difficult time getting in and out. This height also makes it difficult for rescue personnel to get in and pull personnel out from the front or back. They would have to physically enter the vehicle to retrieve people who cannot get out by themselves.

7. Any vehicle used for security details or convoys needs to be able to accomplish several things. First they must be able to push conventional vehicles out of the way, which requires bumper to bumper or frame to frame contact and these vehicles don't provide that. Also, ramming through barricades requires the same thing, which could not be accomplished with these types of vehicle. This type of vehicle has a better chance of running over a standard size vehicle with one wheel and getting high centered, or becoming entangled with its bumper in the upper frame of a standard vehicle. Due to their frame height, these vehicles could actually flip over during such a maneuver.

Not all vehicles of this type of extreme high profile are
built the same. Some are built with the frames lower
and with tires easier to change.

Other questions that should be raised by these types of
vehicles are:

• They are very easy to see from a distance, and
thus they are easier for surveillance to track and for
potential attackers to time for entering an ambush site.

• They look military, and a lot of attacks in high threat
environments are specifically targeted against military
and or law enforcement. Having a vehicle that looks
like either one of these means you could just become
a target because of the threat's misidentification of a
vehicle in your convoy.

Perception is everything. These types of vehicles have
their place and uses but do they meet your needs in
your operating environment?

COMPARING HIGH PROFILE VEHICLES TO OTHER VEHICLES

Compare the bumper heights of these 2 high profile vehicles (above) to an SUV, then compare the SUV to a sedan (left).

Look at the potential striking points those high profile vehicles have if they attempted any defensive driving maneuvers.

CONVOY EQUIPMENT

There is some equipment that will need to be carried in vehicles during movements. Of course, some of the stuff will be required just for the vehicles.

• **Fire extinguishers**. I would have at least 3 in each vehicle. (Remember fire extinguishers are for people first, then equipment and they need to be attached to the frame.)

• **Medical kits** that are stocked for trauma and burns with plenty of IVs. Also if there are any special medications needed for persons on the detail either clients or security personnel ensure that each kit has some.

• Extra tow ropes, tow straps, and other **vehicle recovery equipment** that might be necessary for the mission environment.

• **Heavy duty jacks**, with full size **spare tires**. Also you probably want to carry a 2x2 piece of wood about 1 inch thick in each vehicle for the jack to sit on if you are operating in the desert or in areas with lots of soft dirt, sand or roads that turn to mud.

• A **shovel** in each vehicle is a good idea.

• **Water and meals** that can be stored for long periods. You never know when you might break down and need water and food for the occupants.

• **Sledgehammers and crow bars** might be needed to recover people from vehicles that have been damaged either by the threat or by vehicle accident. Ensure you have the materials available to open doors on any vehicle.

• You might consider a **grappling hook and rope**. If you have cars that are armored like we previously talked about you might need a hook to attach to a door and then use another vehicle to pull it open.

• **Ammunition, flash bangs and smoke grenades** should be placed in each security vehicle. If you have to engage while still in the vehicle you should use the ammunition that is in the vehicle, never use what is carried on your person. This way if you have to exit the vehicle you exit with a full load of ammunition. To store ammunition in vehicles you can attach bags to the back of the seats in front of you or bolt cans on the floor board.

• **Maps and spare batteries** for equipment, such as radios and GPSs.

• **Nontechnical signal devices** such as mirrors, flares and signal panels (VS17) so you still have the ability to signal friendly forces if all your tech fails.

• **Extra belts and fluids** for each vehicle in case something breaks or leaks because in a lot of areas those items for your vehicles might not be available. To plug leaks I also carry epoxy putty and radiator leak stop in the vehicles.

• **Extra cable** for any electronics such as outside GPS antennas or vehicle trackers, and radio antennas. These can be cut or break very easily and having the ability to replace and repair away from your home site is important.

Picking the right vehicle for each mission is just as important as picking the right protective equipment, weapons and people. Having a one size fits all vehicle policy in the security company you use will eventually make you a target, especially if you are always High profile/High visibility. Your best bet for AVOIDING incidents is to be unpredictable in everything you do, even your use of vehicle types and security profiles. The more unpredictable you are the harder it will be to plan for and defeat any security measures you are using in your movements for missions.

CHAPTER 19
CAR OPERATIONS AND TACTICS

Low profile vehicles are one of the best ways to move in high risk areas, because traveling without anyone knowing you have passed by is an excellent way to move, but may not be the best option or the option the client wants to use.

When you do not have the resources and people, a two vehicle movement might be all you can use. These movements can be just as safe as a 10 vehicle convoy with proper planning, training and a good understanding of tactics. The detail leader always rides in the vehicle with the client, where he will navigate and maintain contact with the follow car and observe for suspicious activity. The driver will scan for the safest way to go and give updates to the detail leader as they approach identified critical navigation areas. The clients need to become part of the overall security posture of the movement, with their eyes up scanning, looking for activities, vehicles or people that the detail leader may miss. They especially need to keep their eyes on the roof tops in urban environments.

The follow car will provide security at all stops, and block other vehicles as necessary. The shift leader in the follow car will check navigation, and keep good 360 degree observation going along with the rest of the security detail. 360 degree observation includes not only the follow vehicle but the main body as well. (See page 36 for an illustration of 360 degree observation using 3 vehicles.)

The follow car must maintain distance and speed not allowing any cars to get in between them and the main body. They do not want to follow too closely so they will not rear end the main body adding to the problem if the main body has to conduct an evasive maneuver such as braking,. The main body driver needs to let the follow driver know what he is doing. For example, if he is going left around a vehicle he calls this out so the follow driver can prepare for the same thing. There will be lots of weaving in and out of traffic, passing slower moving vehicles, stopped vehicles, and vehicles turning. The following are suggestions or guides on how you should do some common things while moving.

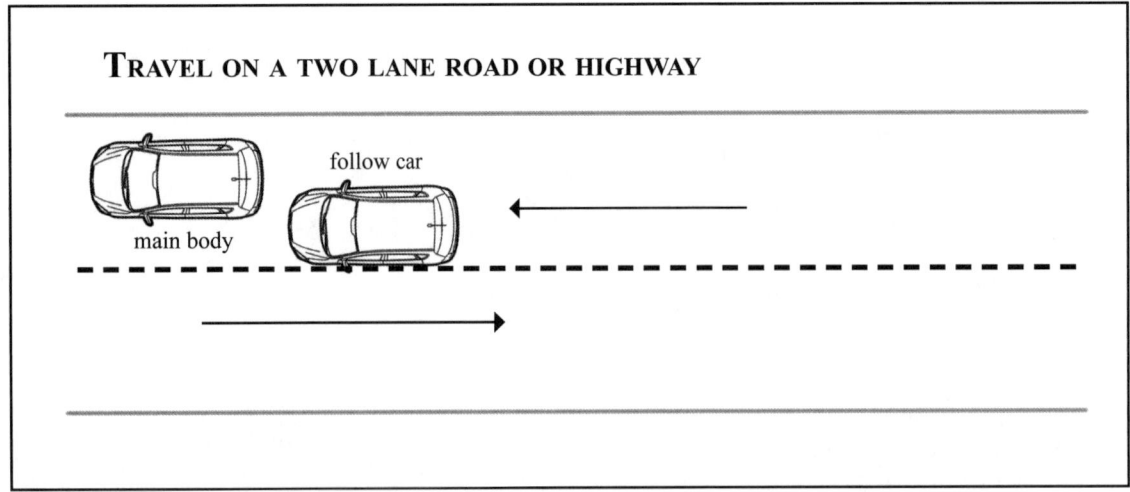

TRAVEL ON A TWO LANE ROAD OR HIGHWAY

follow car

main body

TRAVEL ON A FOUR LANE ROAD OR HIGHWAY

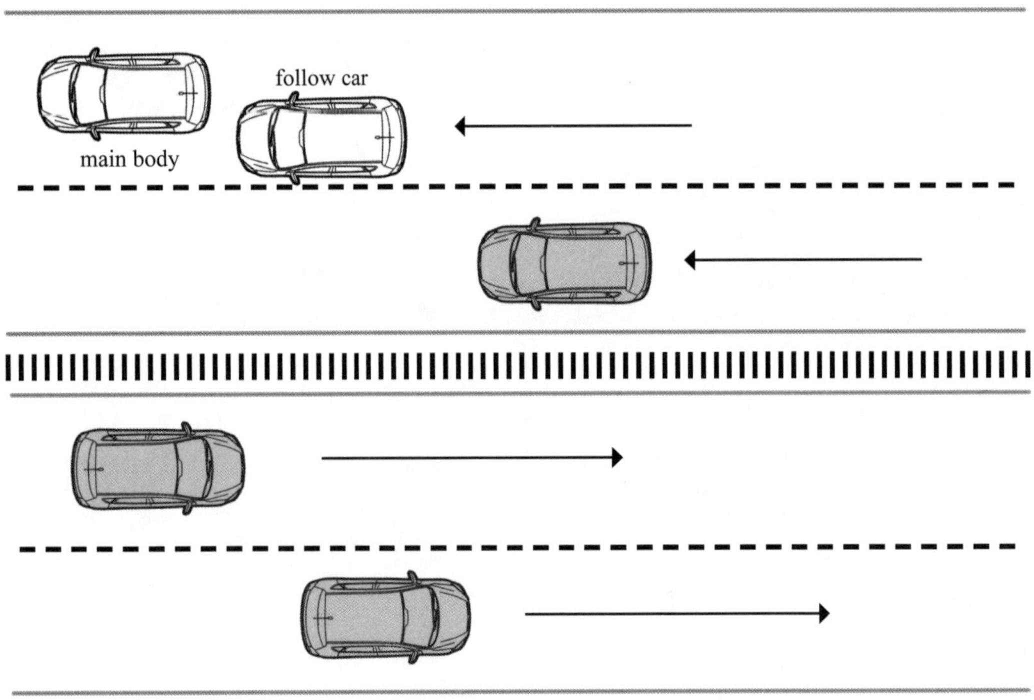

follow car

main body

TRAVEL ON A MULTILANE ROAD OR HIGHWAY

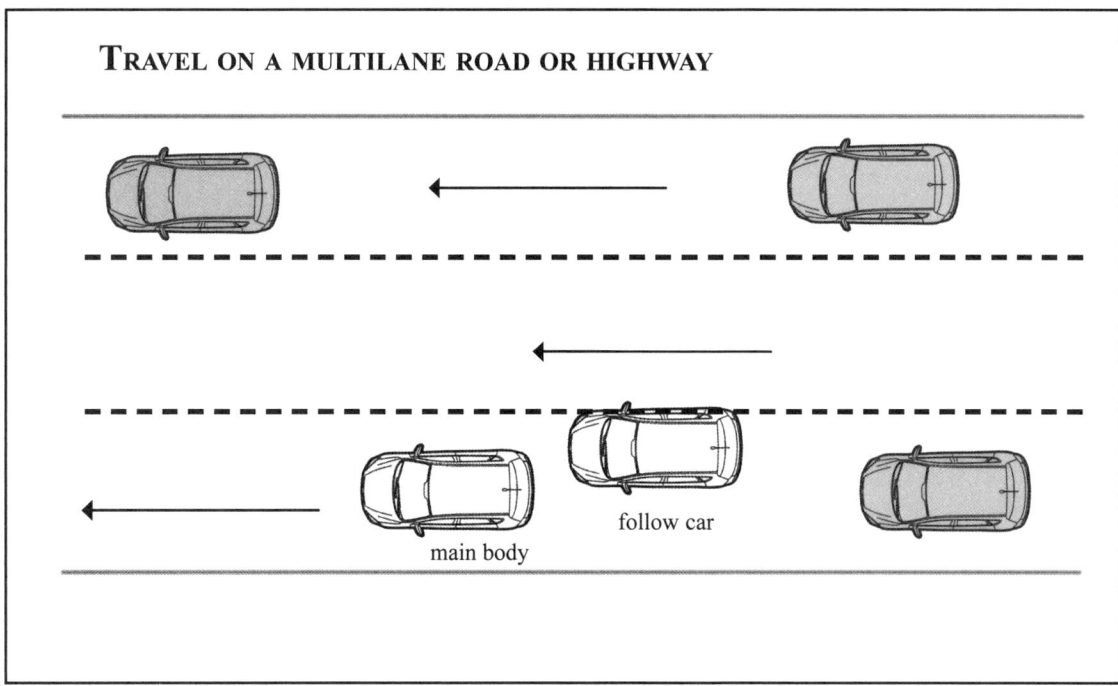

main body

follow car

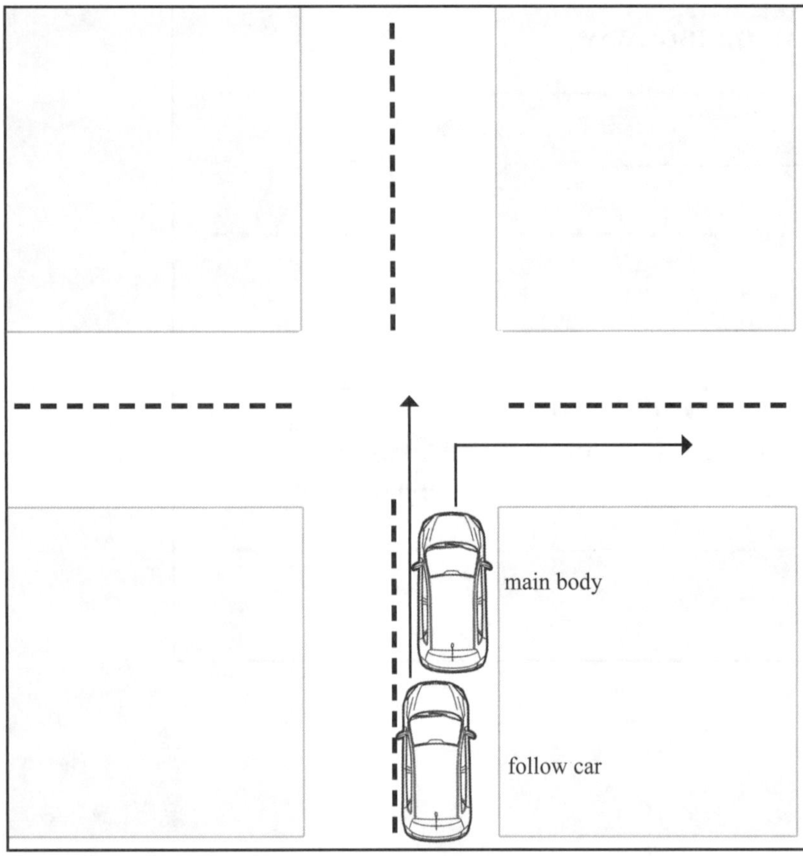

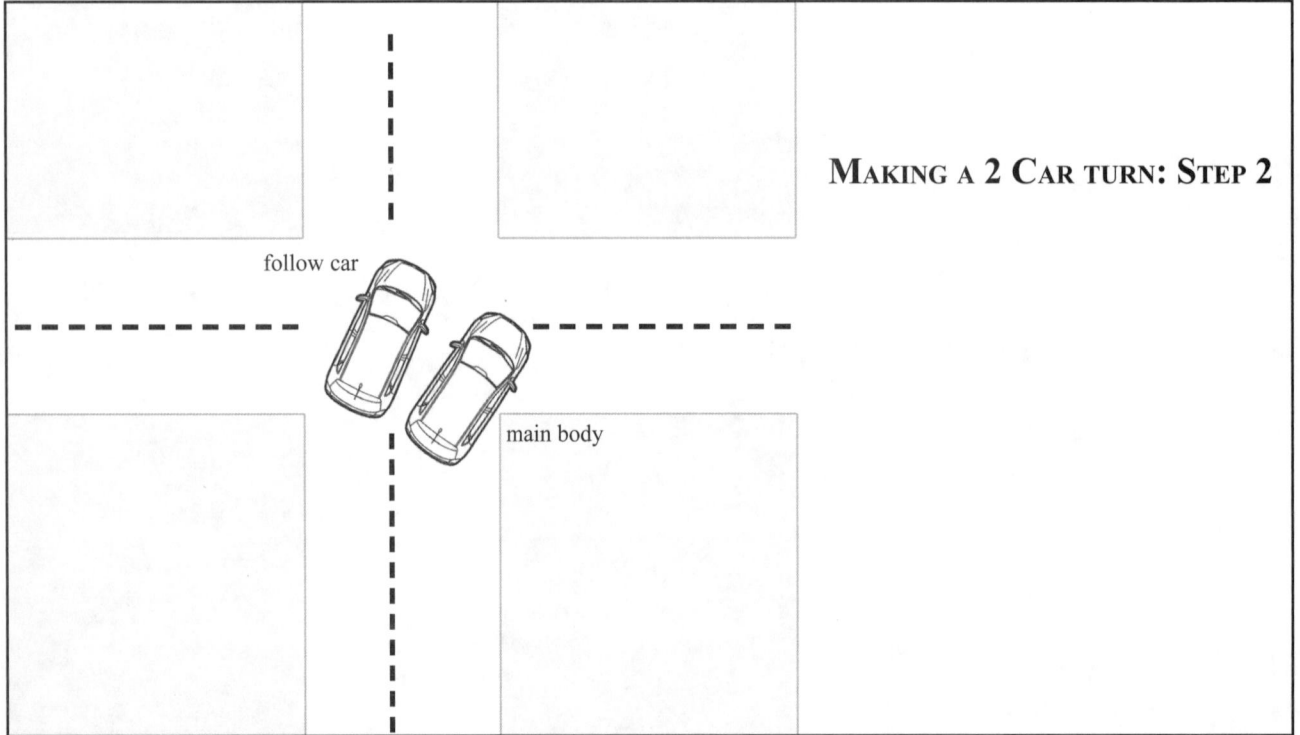

MAKING A 2 CAR TURN: STEP 1

main body

follow car

MAKING A 2 CAR TURN: STEP 2

follow car

main body

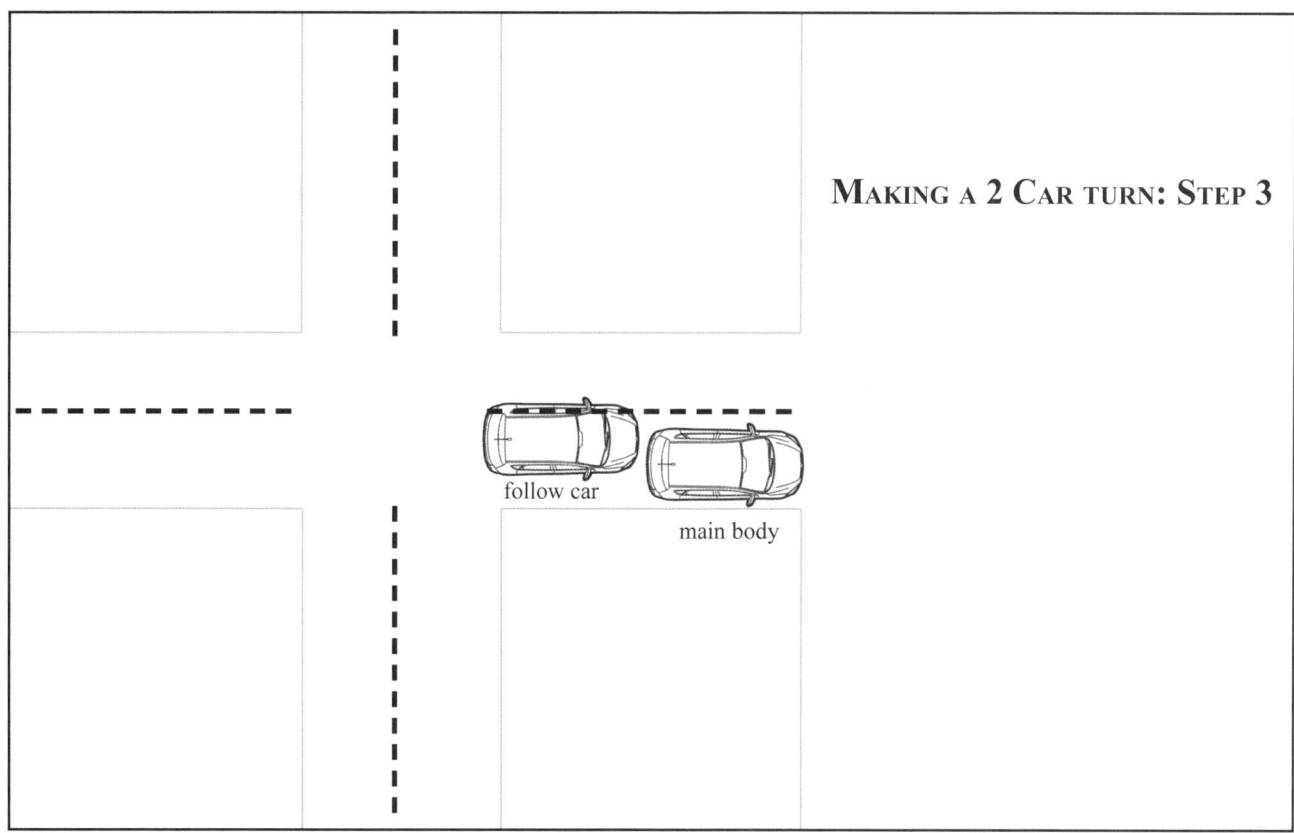

follow car

main body

MAKING A 2 CAR TURN: STEP 3

TRAFFIC CIRCLES

Traffic circles or roundabouts can be very difficult to negotiate at the best of times, and even more so when trying to maintain a 360 degree perimeter. Since most traffic circles do not allow parking inside the perimeter, the follow car or lead car needs to look at slowing and stopping oncoming traffic because vehicles heading at your client at speed are the biggest threat. Its best to keep use of traffic circles to a minimum and when doing your routes plan on staying in the circle for no more than 1 turn off. Get in and get out to your required turn as soon as possible!

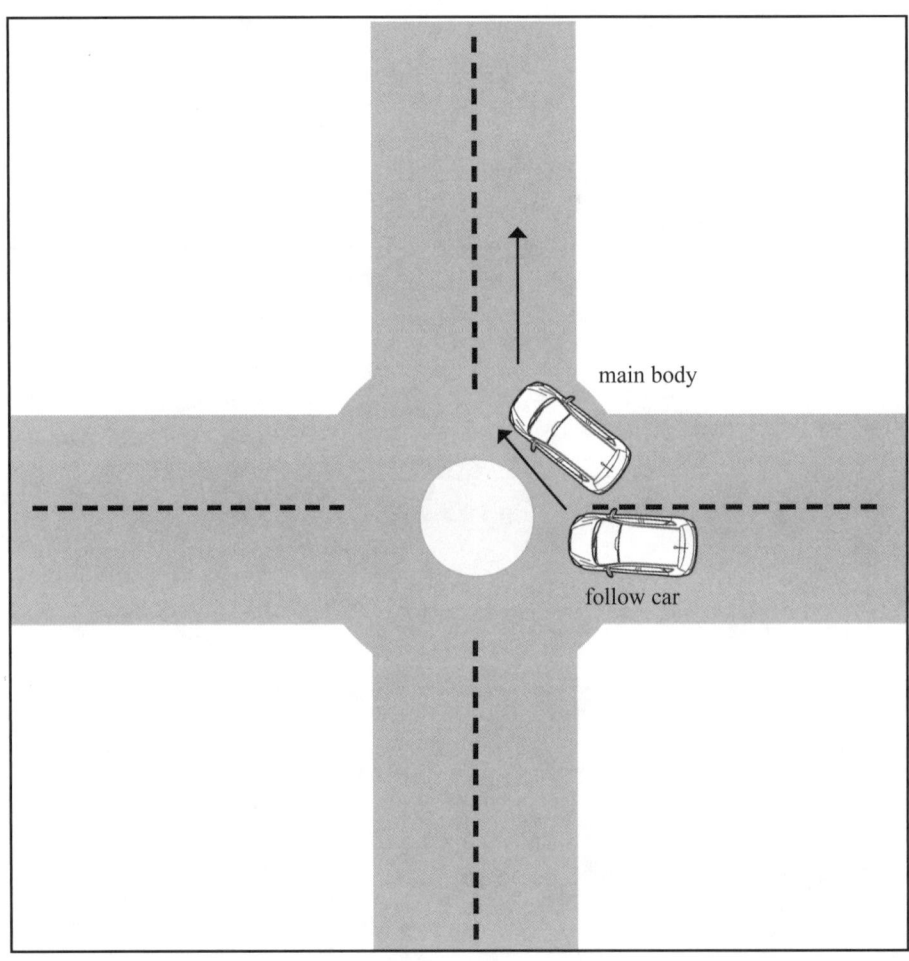

main body

follow car

VEHICLE RECEPTION OF 2 CAR MOTORCADE

Arrivals of the client are one of the most dangerous times since he is now place predictable especially if it is an advertised event or function. With a two car convoy it is important to make sure that at least three advance people are there to meet the client when he arrives to help set up the security bubble. If you have no advance make sure that the event security or static security at the facility is there observing and providing perimeter security. If at all possible do your arrivals in covered or underground parking facilities that have restricted access. This may not always be possible if your client needs to be seen arriving by the media or for other business reasons or because of the layout of the facility. Everyone, upon arrival, becomes a part of the security bubble even it is only for observation. Drivers will observe their designated areas of observation and security personnel will exit the vehicle and take up designated positions around the perimeter close to the vehicles. Vehicle doors should be closed after people exit to reestablish the armor seal and protect the drivers from incoming rounds if an incident occurs. It doesn't take that long to open a door.

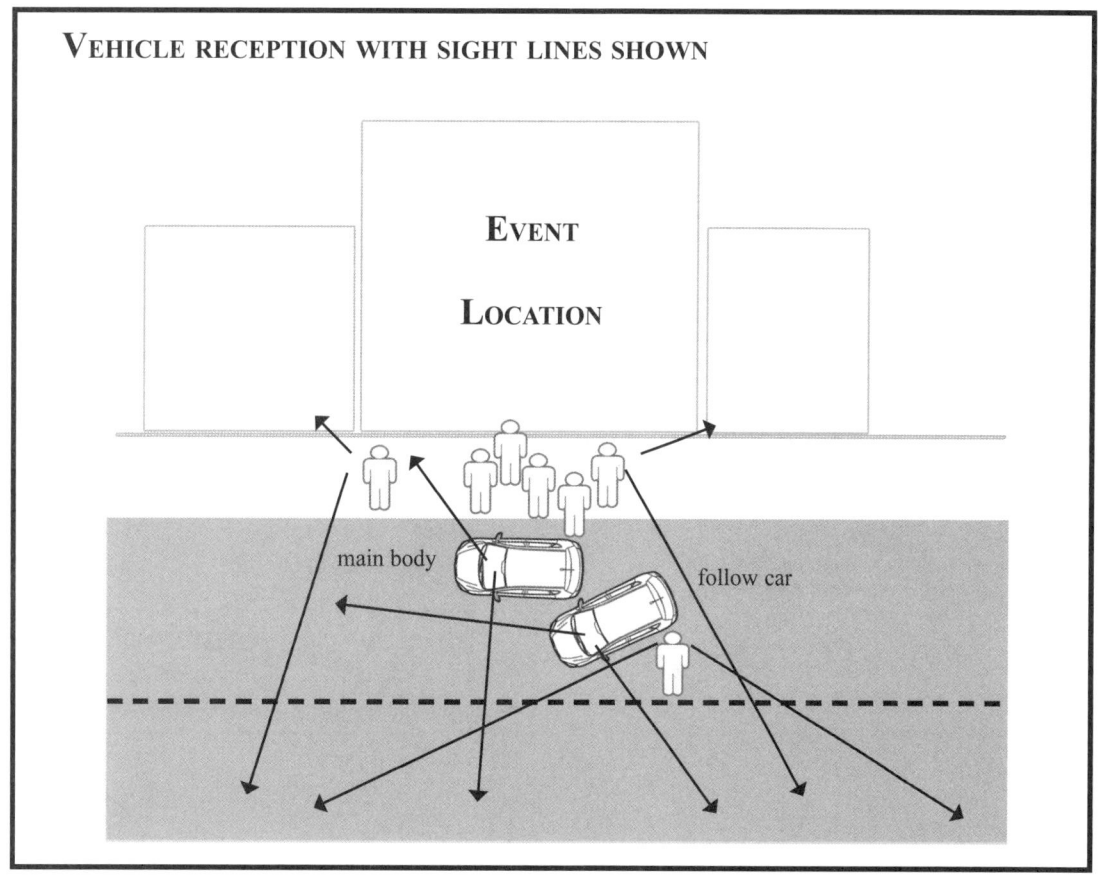

VEHICLE RECEPTION WITH SIGHT LINES SHOWN

EVENT

LOCATION

main body

follow car

2 CAR MOVING ROADBLOCK ESCAPE

Moving roadblocks are used to slow and stop vehicles so the threat can disable them, take out security, kidnap or assassinate the client. The most effective way for a 2 car detail to deal with this is for the follow car to bypass the main body, and pit the car on the left of the main body or strike it forcing it up and away from the main body vehicle. The main body vehicle will then stop, and if the area is clear, do a reverse 180 turn. If not clear he will execute a fast 3 point turn to get the vehicle turned around and moving away from the threat vehicles to a safe area. The security vehicle will disable the threat vehicles if possible then follow the main body to remake the detail and reestablish security. License plates and details need to be remembered after such an incident for the police report. It is better to have a dash mounted video camera and digital cameras available in the vehicles to take pictures during incidents for better identification of the threat to police and intelligence agencies.

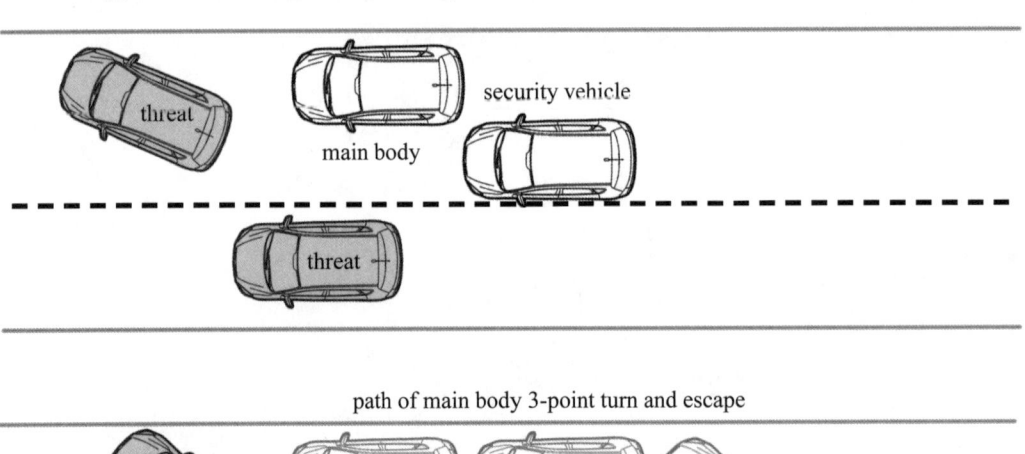

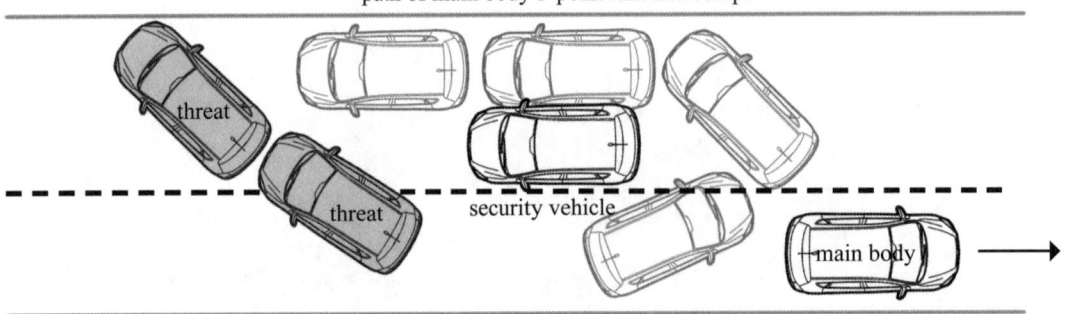

2 CAR STATIONARY ROAD BLOCK

Stationary roadblocks are a problem no matter how many vehicles you have, if you don't get enough warning to make a decision, such as turn around and go another direction. For a movement with two cars, the follow car must pull in front of the main body vehicle and block the stationary road block, while the main body gets out of the area and heads towards a safe haven. If there is any attempted pursuit the shift leader and the security detail in the follow car must deal with the threat to allow the main body vehicle to put as much distance as possible between the client and the threat. The follow vehicle, if it is able, will rejoin the main body at a safe area, and they will find an alternative route to their destination.

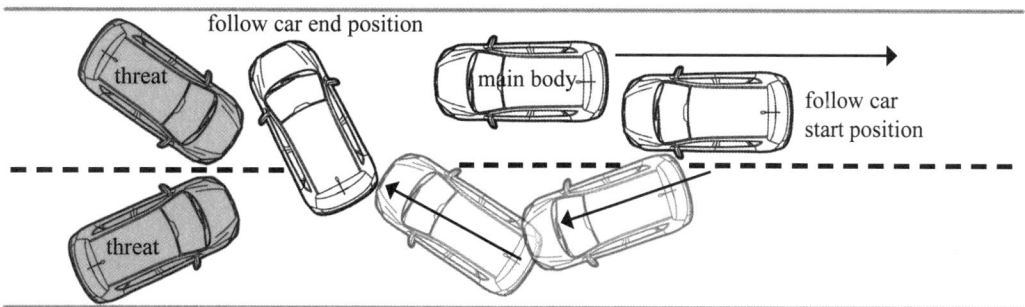

3 CAR OPERATIONS AND TACTICS

3 car details are what should be used as your standard for most operations. Remember, your detail size will depend on the number of clients you are moving, the risk you have to manage, and the threat as well as whether you are going low profile or high profile, so no detail mission will ever be exactly the same. 3 car missions can be used any time but are mostly used in permissive and semi-permissive environments when you have some type of local law enforcement presence in the area, security at the sites that will be visited, and good intelligence on routes to be used.

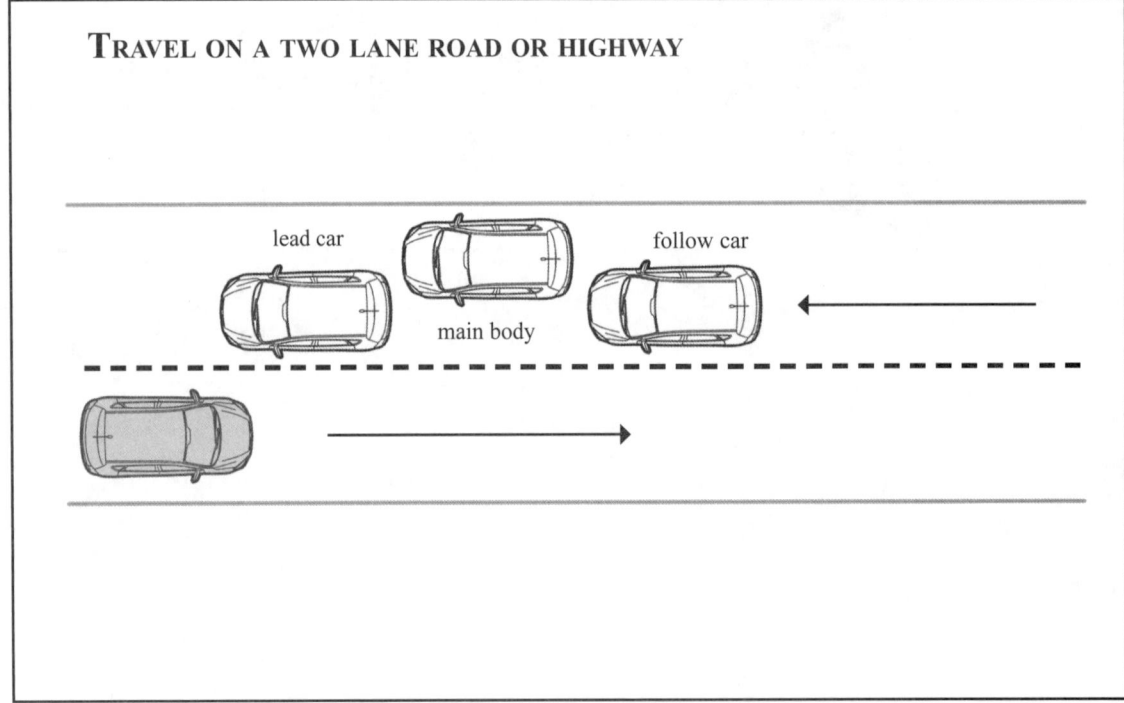

TRAVEL ON A TWO LANE ROAD OR HIGHWAY

lead car

main body

follow car

Travel on a Four Lane Road or Highway

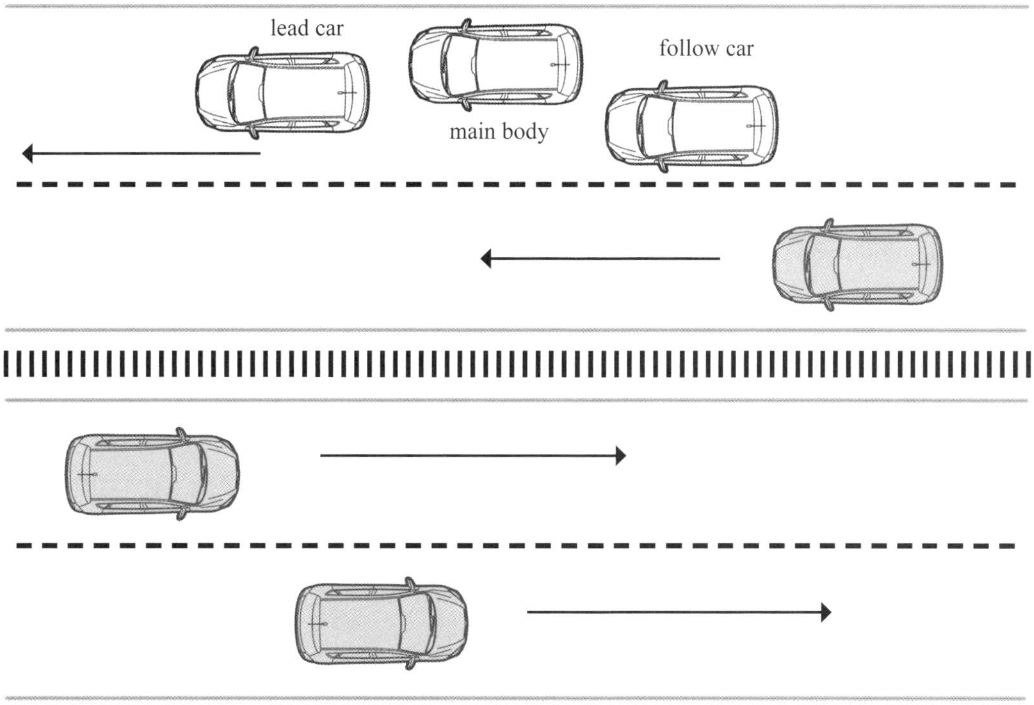

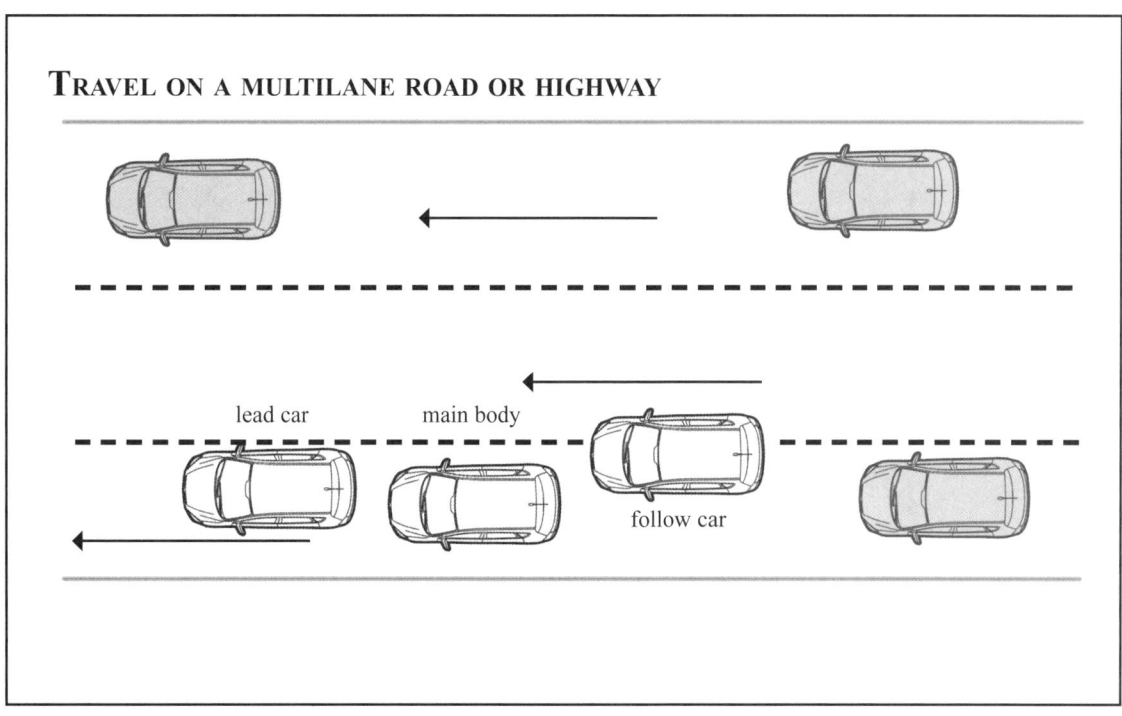

Travel on a Multilane Road or Highway

3 CAR TURNS

Turns with 3 or more vehicles will basically be all the same. The car immediately in front of the main body vehicle will move in and block the intersection, as the main body vehicle turns. The lead vehicle will then move back to the front of the main body as the follow car replaces it blocking the intersection, then the follow car will fall back into position as the main body passes. This technique is for all intersections including "T" intersections and protects the main body vehicle at its most vulnerable point, when slowing to make a turn.

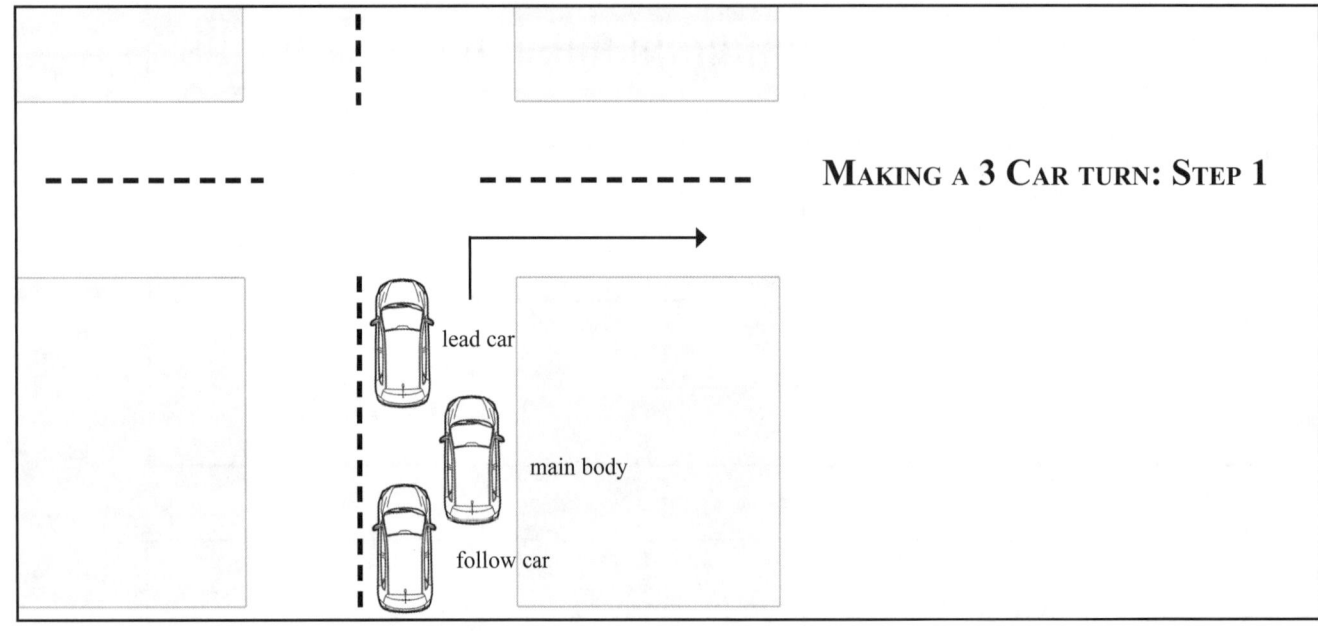

MAKING A 3 CAR TURN: STEP 1

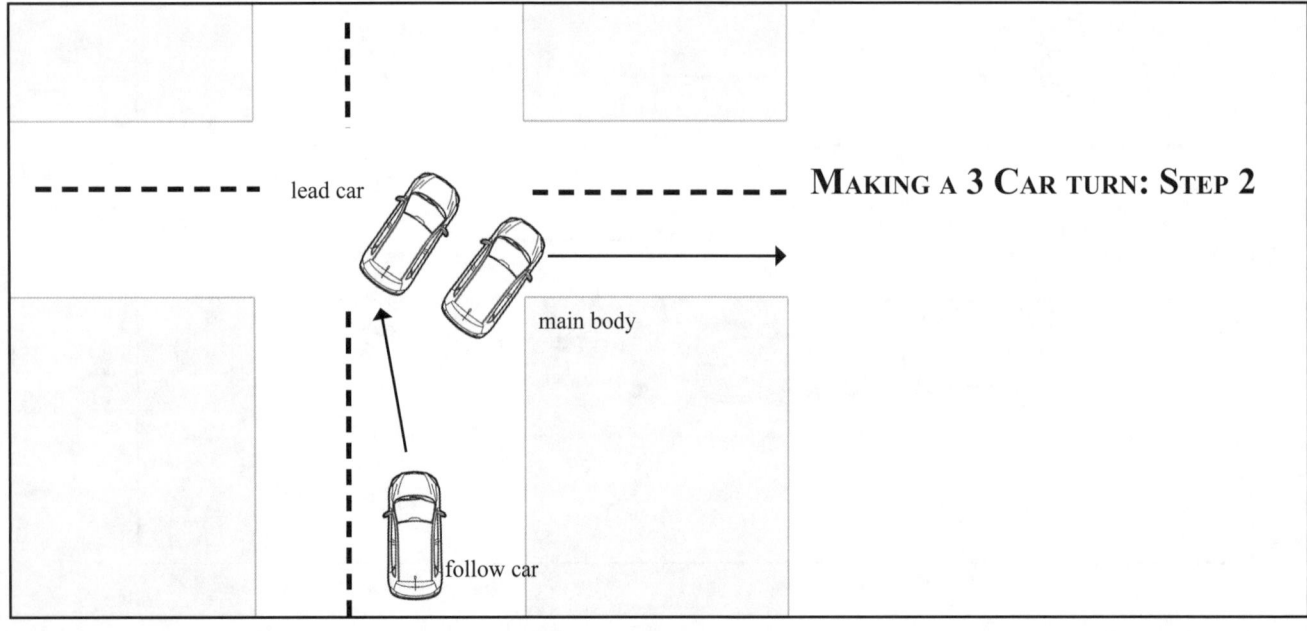

MAKING A 3 CAR TURN: STEP 2

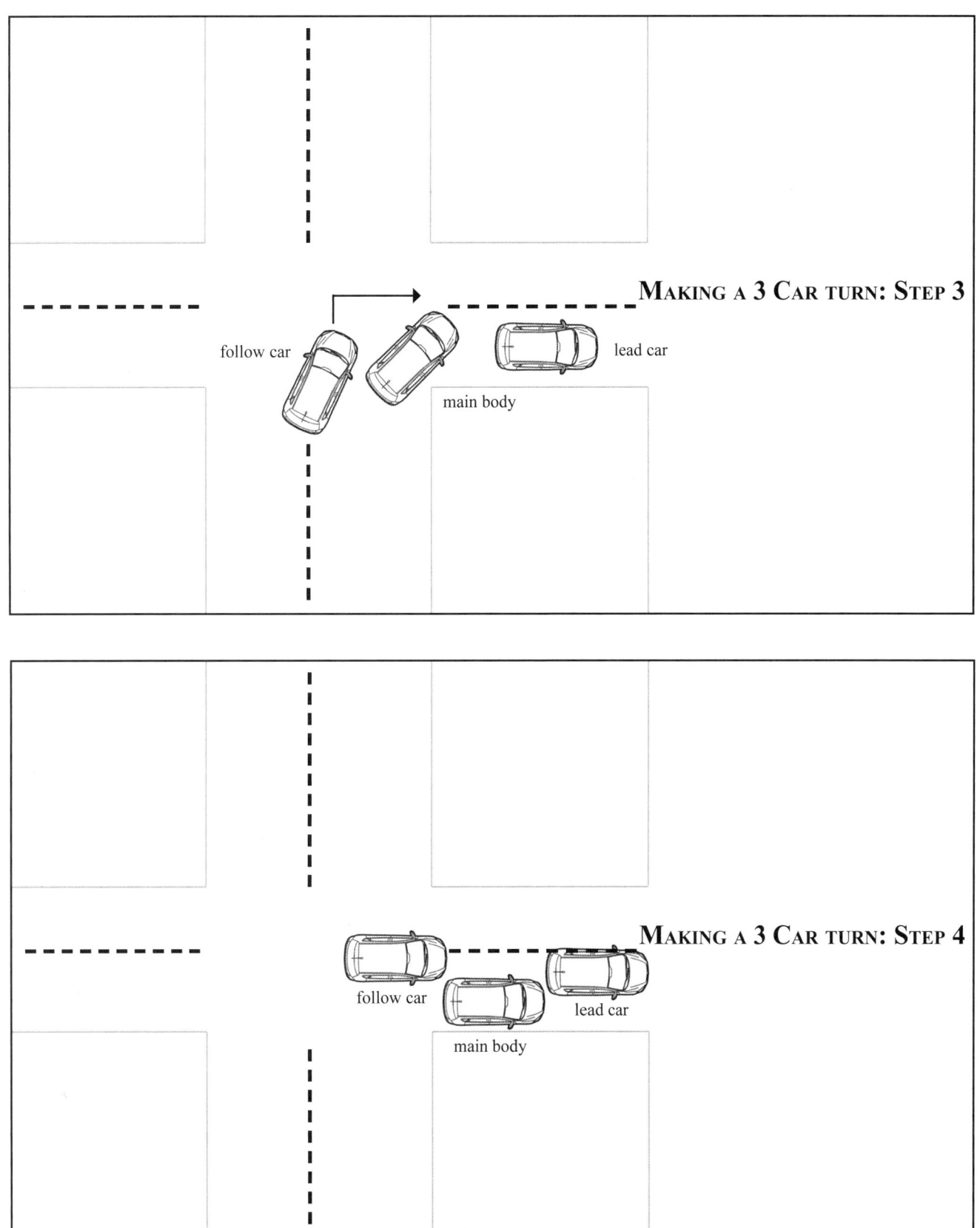

MAKING A 3 CAR TURN: STEP 3

follow car

lead car

main body

MAKING A 3 CAR TURN: STEP 4

follow car

lead car

main body

LANE CHANGES

Lane changes will be the same when working with 3 or more vehicles. The follow car will move over and block the lane the convoy wants to move into then the main body and the lead vehicle move over into that lane. This technique allows the security detail to control the security bubble around the main body vehicle as it moves in traffic from lane to lane and allows the whole convoy to move at one time, staying in formation.

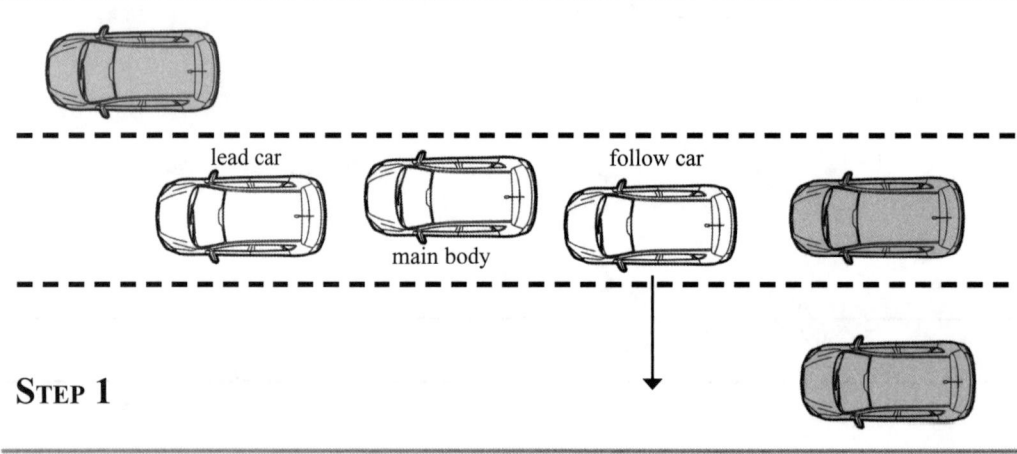

STEP 1

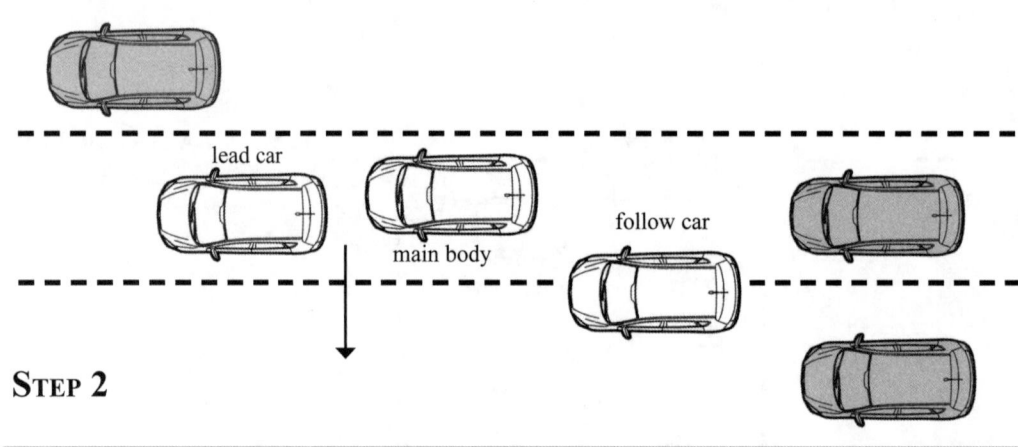

STEP 2

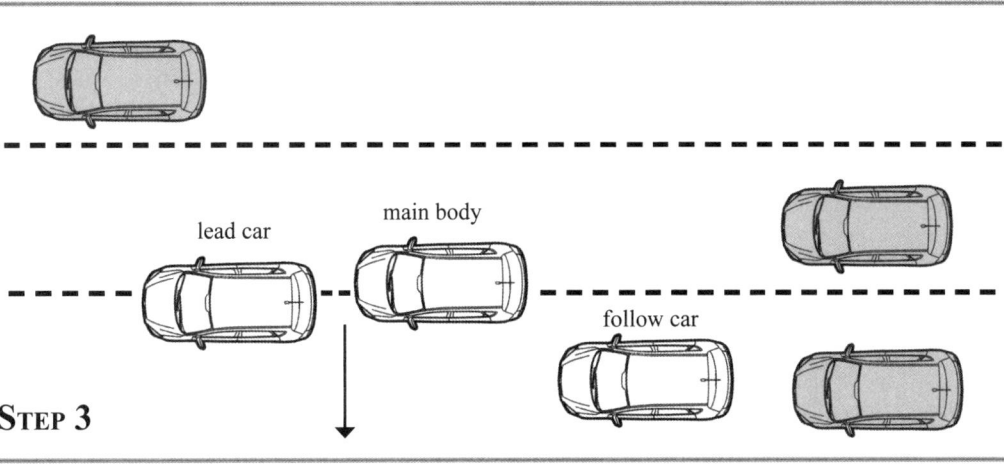

STEP 3

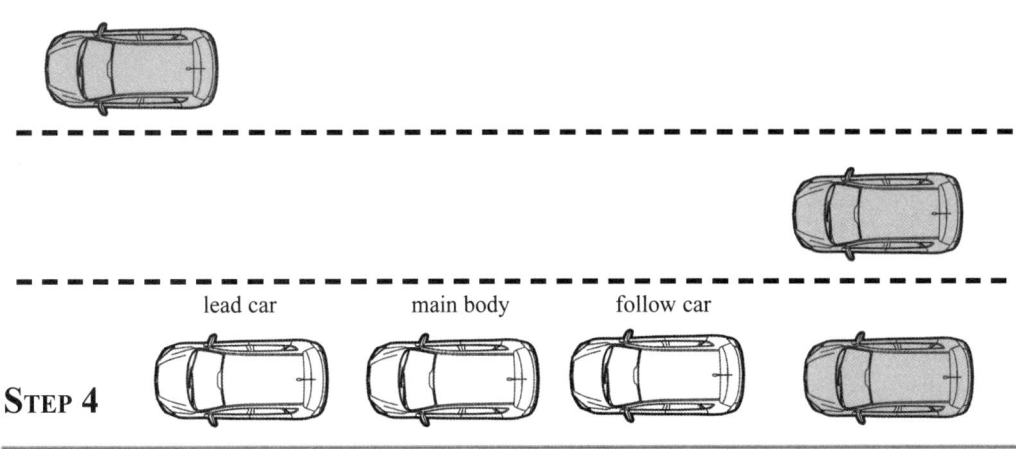

STEP 4

APPROACHING AND PASSING THROUGH INTERSECTIONS

When a convoy approaches an intersection to drive through it, like all other traffic control points, they need to control the vehicle movement in and around the main body vehicle. The lead vehicle will pull up and block the intersection to the left, while the main body will slow down, moving to the center of the lane or road and the follow vehicle will pass it and block the intersection to the right. As the main body passes throughthe intersection, the lead vehicle will speed up and resume its place then as the main body passes the follow vehicle it will fall in behind and resume its normal position in the movement.

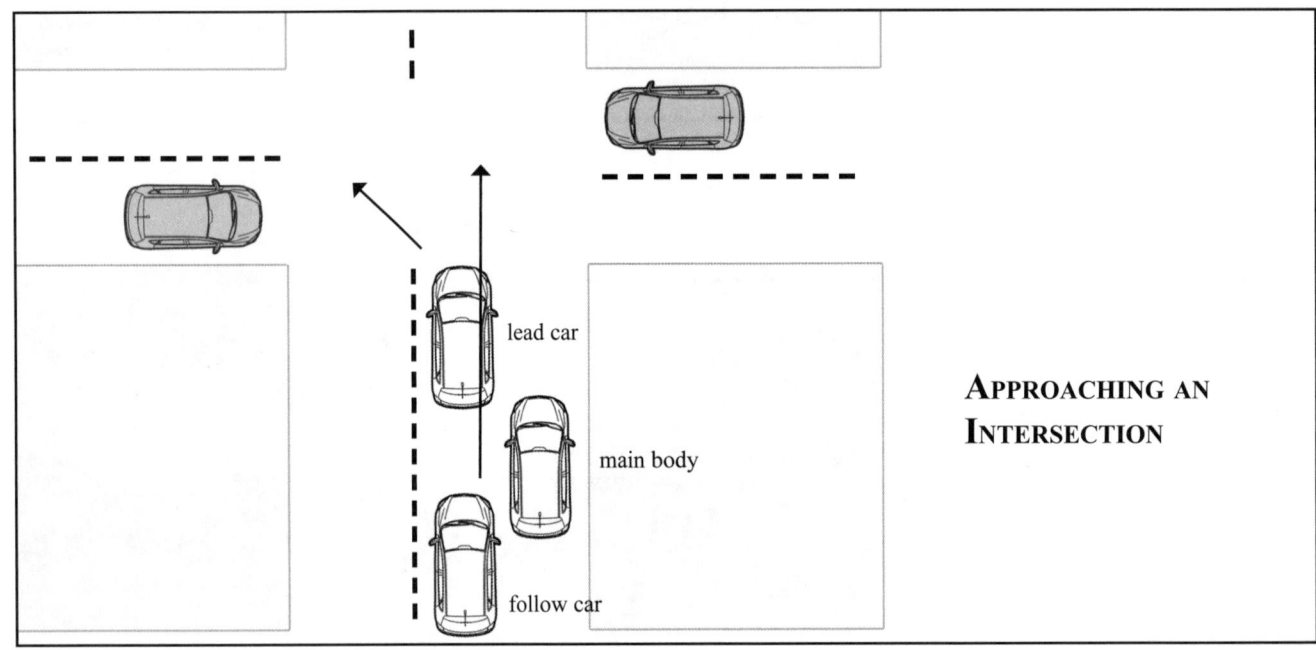

APPROACHING AN INTERSECTION

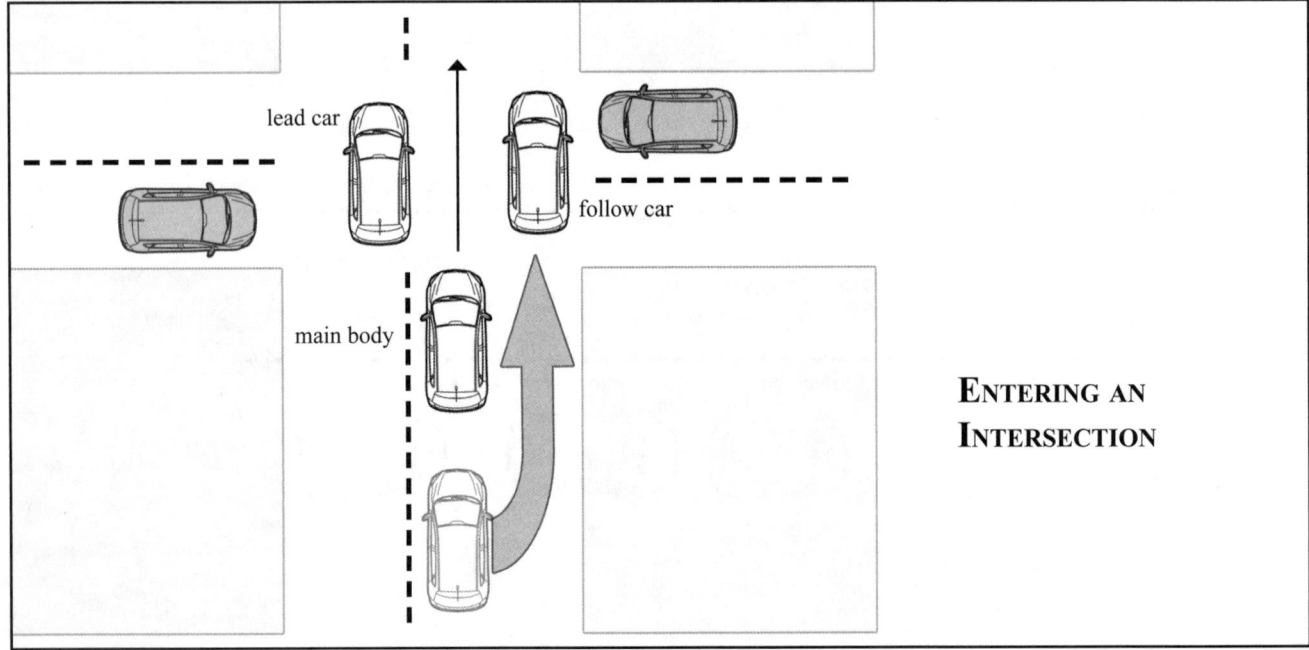

ENTERING AN INTERSECTION

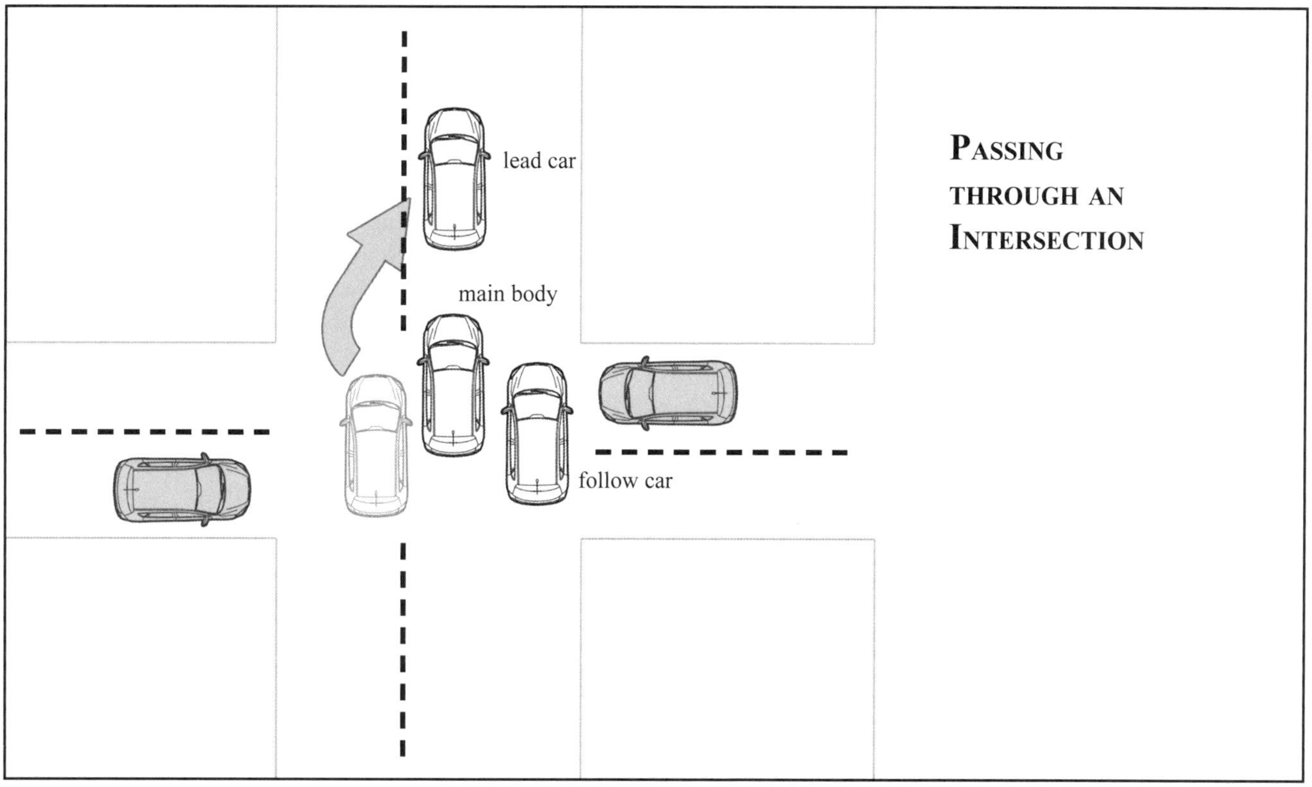

PASSING
THROUGH AN
INTERSECTION

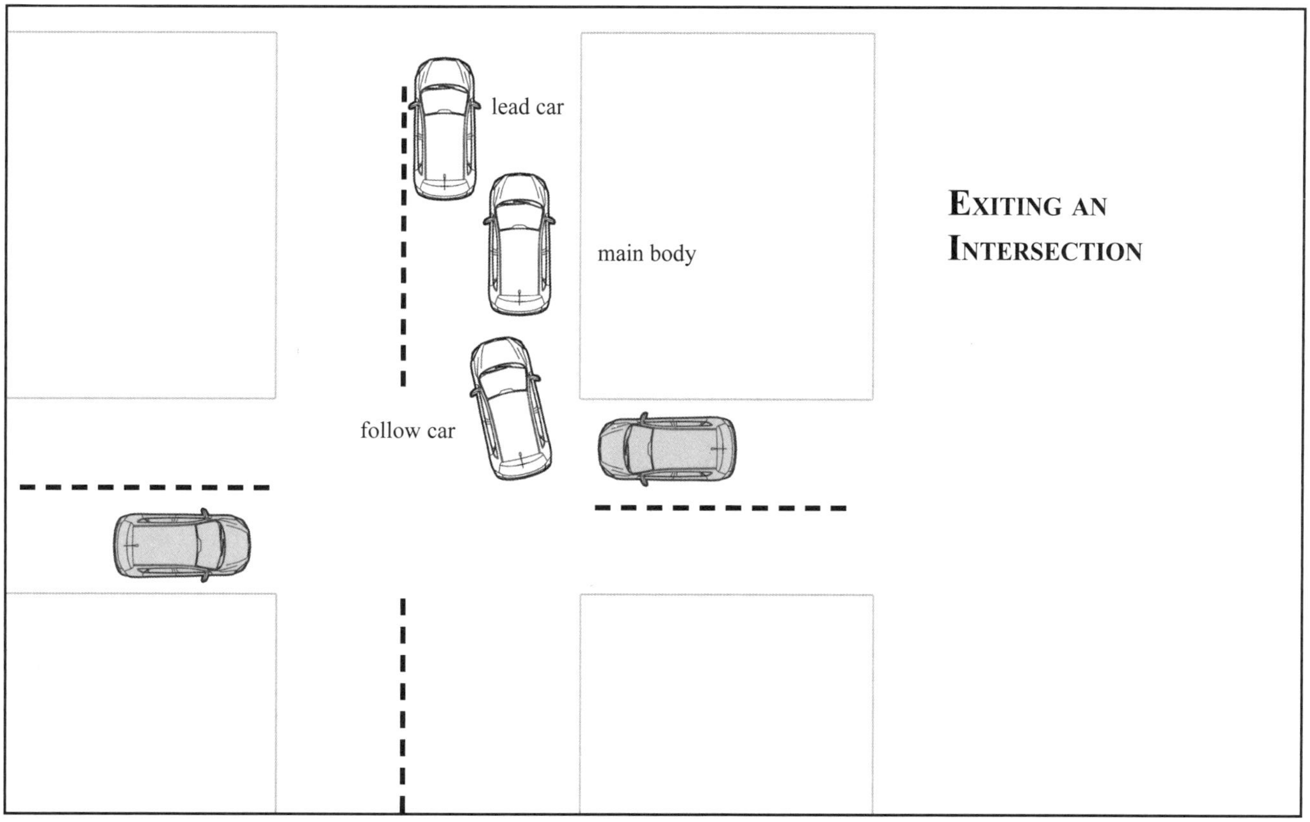

lead car

main body

follow car

EXITING AN
INTERSECTION

lead car

main body

follow car

VEHICLE RECEPTION 3 CAR MOTORCADE

Vehicle receptions for convoys with 3 or more vehicles will basically always be the same. The advance or site survey should have told you how the approach will be done and how much maneuvering space you will have. The lead vehicle will pull in, leaving enough space between it and the curb so that the main body or main bodies can drive right next to the curb. The follow vehicle will then pull in and park on the outside of the main body. When pulling up, the main body will pull up alongside the lead vehicle as far as possible without blocking the rear door, then the follow vehicle will pull up beside the main body far enough to not block its front door. The detail will then exit setting up a 360 degree area of observation and protection.

This creates a wall of protection for the client as he exits and is outside of the main body vehicle and still allows all of the doors to be used by the security detail. This also means all vehicles have unrestricted movement forward or back without being blocked in by one of the security vehicles. If an advance is there then they should already have static security in place with the advance leader meeting the client and moving him inside quickly. If there is no advance, the client will stay in the vehicle until the security detail sets up its security perimeter, then the detail leader or bodyguard will move the client into the facility. All doors to the vehicles on the exit side will remain open until the client is safely inside the facility, and all vehicle doors on the outside will be closed as the security detail member exits to take up his position.

This will be duplicated when the client leaves, setting up the area or zone of protection, and having the vehicles in place before he exits the building. Drivers remain in their vehicles at all times with doors closed.

There may be more than 3 vehicles because sometimes a convoy may have more than one main body depending on the number of clients being moved. Also if a convoy has a security advance patrol vehicle, or a counter assault vehicle some companies make them part of the embus or debus formation. I prefer to only have the lead, main body and follow vehicles as part of this formation and keep any extra security back until they are needed. This hopefully keeps them as an unknown part of the detail and they can react directly to any incident, while the rest of the detail gets the client to safety. If they are used as part of the embus and debus procedures they will become a known part of the security profile and the threat can anticipate them and then plan for them as a known fact instead of them being an unknown factor in the equation.

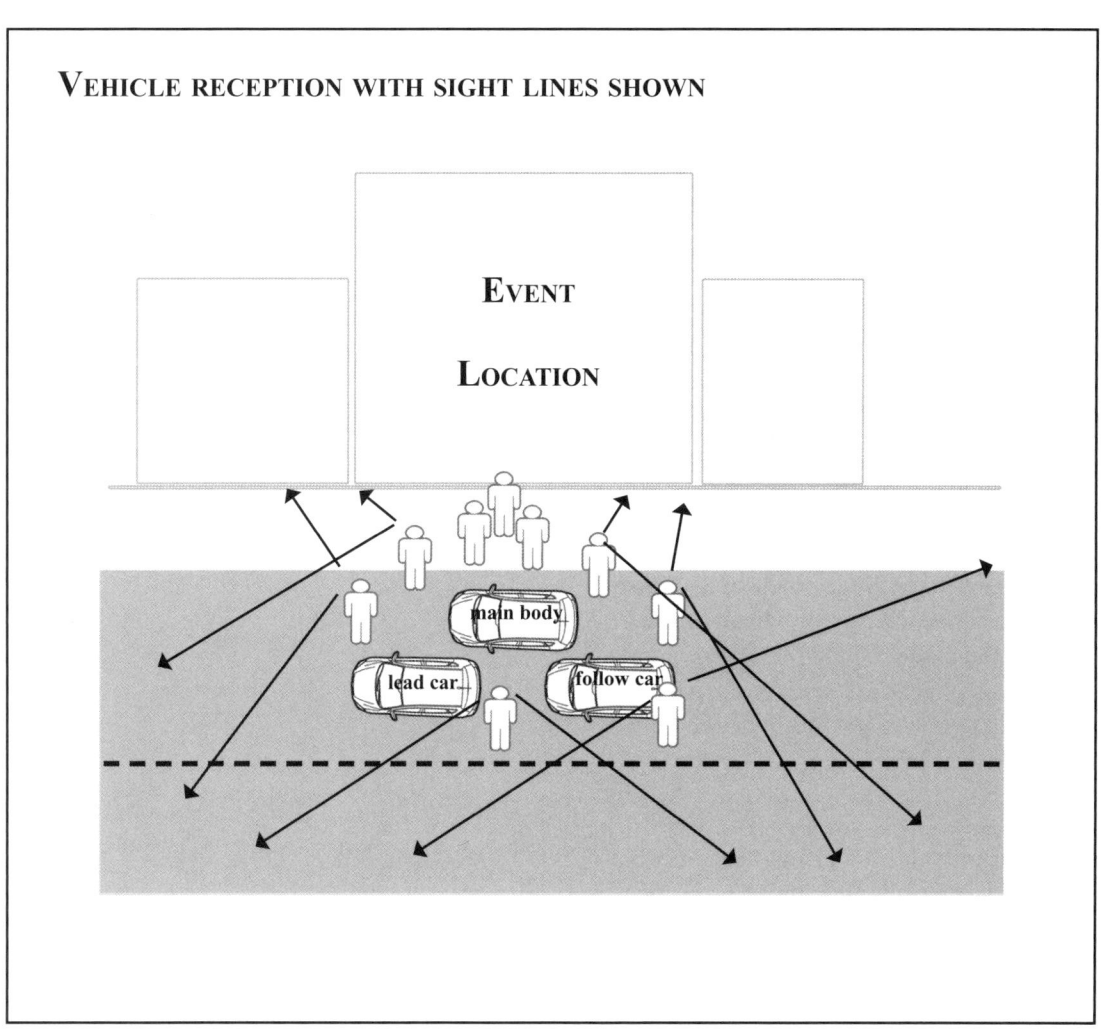

Vehicle reception with sight lines shown

REACTING TO ATTACKS AT THE EMBUS/DEBUS POINT

When the client is moving to get into the vehicle (embus), the response depends on his location. If he is closer to the building, the shift leader or bodyguard (BG) will lower the client's profile, cover him and move him back inside the building. If he is closer to the main body vehicle, then the detail leader will lower the client's profile, cover him and put him in the vehicle. The detail leader or BG should be the one making the decision on where to go, not the client! As the most experienced and highly trained person, the detail leader will physically control the client if necessary and get him to the closest cover.

The vehicle will immediately evacuate the area. Once the client is evacuated then the security detail will load up and link up with the main body to provide mobile protection until the client gets to a safe haven. The security detail should not be staying to fight once the client is gone, its not their job.

When the client is in the vehicle and an attack occurs at the debus site, the vehicle along with the security convoy will immediately evacuate the area. Everything else is basically the same. The detail leader or BG will move the client to the closest cover. If it's the vehicle, they will immediately evacuate. If it is the building, get the client inside to a safe area and wait to be evacuated.

These are just a few general options. You need to come up with your options and courses of action for each site, and practice them. There are many different ways to do things. Come up with one that works for your locations, experience, resources and environment and make it work. Just remember, you have to do something. Doing nothing will get the client and you killed.

CHAPTER 20
THE COUNTER ATTACK TEAM (CAT)

There will be times and places when a basic security detail will not be enough to guarantee the safety of a client. During situations like this added security might be needed to give that extra layer of protection to protect the clients and have a successful operation. This extra layer of protection can include 3 things: a Security Advance Patrol (SAP), a Counter Attack Team (CAT), or a Quick Reaction Force (QRF). The detail leader for the mission will make the decision based upon the threat and risk assessments.

These details are basically a specially trained quick reaction force deployed in support of a personal security detail during its movement and at any static sites. It needs to be capable of carrying out a fast, aggressive response to any attack on the detail or client, with little or no direct control from the detail leader. Most CAT/SAPs will consist of 10 men in two vehicles but this is a decision that each company will make. Some have both vehicles behind the convoy, others use one vehicle in front as a SAP and one in the rear as a CAT. Environment, experience, and routes will help you make the decision on what is best.

The basic principles for a CAT/SAP or quick reaction force are simple: You need a group of individuals who train as a team and are dedicated to this one job. They have to understand that their job is to engage any attackers, draw their fire away from the client and neutralize the threat. The CAT can do this in many ways and should be very flexible in their tactics so they can adapt to any security situation or mission as required. They need to be able to react to an incident as quickly and safely as possible. After getting the call to engage, they need to get to the site, deploy, assess the situation and act with aggressive behavior and accurate firepower. This will help the security detail regain initiative so they can do what is necessary to safeguard the client.

The type of visible profile the CAT/SAP will use needs to be well thought out based upon the current mission, operating environment, threat/risk assessment and local area knowledge gained from the advance and the route surveys. Other considerations deal with the mission itself such as:

- the type of event and its location

- the client's image and what image he wants to project

- the resources and manpower available

With proper planning and forethought there should never be any reason a CAT/SAP could not used if the situation requires one. There are two types of profiles that can be used in any situation for every detail: high or low visibility and overt or covert profile.

High Visibility or Overt Profile. The high visibility profile achieves one aim of security by being obvious. It is there and displaying its capabilities. This is a visible deterrent to any threat surveillance because it raises the security profile of the detail and client by displaying a highly determined and aggressive capability to meet any threat.

Low visibility or Covert profile. The low visibility profile is used when you do not want the threat to know your capabilities. If they do not know about it, they cannot plan for it. The principles of use and capabilities of the CAT/SAP do not change when working in this profile. The CAT/SAP both could be used in a counter surveillance when necessary to increase the security of the movement overall.

The CAT can be either mobile or static with separate methods of operation:

MOBILE ROLE

The CAT will operate in vehicles that meet their needs, in some places that means four wheel drive vehicles and in others vans. They must travel close enough to the convoy to provide a fast response to an incident but not so close they get caught in the incident as it happens. The distance will depend on traffic density, terrain, weather, and the threat. The CAT assists the detail during a threat contact by providing heavy fire support from a location outside the kill zone.

In the event of an ambush on the detail convoy, the CAT will attempt to gain the upper hand by firing and maneuvering on the threat to win the engagement. While they are doing this and drawing the attention of the threat away from the convoy, the security detail will evacuate the clients to the nearest safe haven if possible but they will do everything they can to get the client off the "X". If the incident is an explosive ambush with no small arms fire the CAT will assist the convoy by taking control of the area around it when they arrive and provide a security bubble for the detail to work in. They will assist with medical help if the detail is unable to provide it, and they can help evacuate the detail and client as necessary.

The CAT can also be used in case of a traffic accident or vehicle break down, by helping set up a security perimeter until the situation is dealt with. If they are using a high visibility profile then you can use them at danger areas and choke points and at the site location at the embus/debus areas as a show of force. This show of force could deter a possible attack, by presenting an unknown security factor. You can use the team also by having it travel in front of the security convoy by several minutes to scout out the route for any problems and as a show of force. When used this way it is called a SAP.

STATIC ROLE

When used at a static site, either the residence area or a site the clients are visiting or a work site, the CAT (or now you could call it a quick reaction force) will increase the level of protection provided by the static security at those locations. The CAT/SAP/ quick reaction force can be used several different ways at static locations depending on the working environment, the threat, location of the static site, and the duration of stay at static location. It can be used as a show of force, when deployed with their vehicles to static locations outside a site. They can conduct foot or vehicle patrols around the location and carry heavier weapons than normal security for the area, which can act as a visible deterrent to any threat considering the site for some type of attack.

This detail can also be used as a quick reaction force during day to day operations at a static location, staging out of a facility or building on the premises. I always preferred to have them staged in buildings away from the main building so they would be engaging a threat from behind instead of trying to fight their way out. In this function they can also perform other duties to enhance security operations at a static location at the direction of the site security manager.

The detail can also be used in the counter sniper capacity when the situation and potential threat requires it. They would be deployed in teams to observe and cover positions that a sniper or someone with another type of stand off weapon could occupy and use to attack the static location. They should also cover all routes to the static location including pedestrian and vehicle approaches. Usually they are used in this role for short term operations. You can also use them in this role at choke points and other danger areas along the routes of travel.

The training of the CAT/SAP should concentrate specifically for its role as a shock force during an incident. This includes training in close quarter battle and combat marksmanship, infantry skills of fire and maneuver as a squad to overcome the threat, the security detail's convoy procedures and immediate action drills. They should also train with the security detail and static security guards as often as possible so they know their SOPs, security and defensive driving skills for getting into positions to support the security convoy when necessary, as well as sniper and counter sniper training, and advanced medical training and calling for medevac procedures to help support the security detail after an incident.

When used in the counter surveillance role the CAT/SAP will be in low profile mode, so it will have low visibility equipment including the type of vehicle, weapons, and most especially, dress since they won't be in full body armor, helmets and assault weapons. These necessary limitations for conducting counter surveillance lower their counter attack capabilities, but this is an excellent trade off if they can spot surveillance or an attack forming up allowing the convoy to avoid it.

MOBILE OPERATIONS PROCEDURES

The majority of attacks against a client will come during movement operations, because the threat will have more control over the environment then you do and they can pick where they want to conduct an attack and gain the advantage. A CAT gives another level of security to movement operations and enhances the security provided to the client. A CAT can either be behind the security convoy or in front where it is called a SAP. Both operate under the same principles or procedures.

The type of vehicles used by the CAT will depend on whether it is going to be high or low profile, and this depends on how the convoy will be moving and the threats in the area. A CAT vehicle should have 5 doors and tinted windows or shades. It should be an SUV, 4 wheel drive vehicle with the first passenger row of seats taken so no one has to climb over anything to get out and equipment can be laid out inside. The rear seat should be turned so it faces the back door. If it is a high visibility vehicle use the bumpers should be rigged with old tires for vehicle ramming operations, and it should be pre-rigged for towing operations. Low profile vehicles should not have modifications visible on the outside. When using a low profile vehicle like a car, the CAT or SAP might have to go to 3 vehicles instead of the normal 2 to carry all ten members or go with a smaller contingent. Preplanning will decide which is the better option for the operation.

Weapons carried by the CAT will be determined by the type of working profile, or I should say how they are carried, overt or covert, because for all movements you will need a light machine gun, grenade launcher, flash bangs, smoke, CS grenades, and sniper rifles, along with plenty of ammunition for each weapon, plus extra ammunition in case the security detail needs it.

Other items to be carried by the CAT include rescue equipment for opening a car after a vehicle accident such as seat belt cutters, hack saws, and sledge hammers, large medical packs stocked to handle any trauma needs during movement, extra food and water, and any special items needed to handle foreseen and unforeseen incidents during movement.

The CAT element can be used in several different ways based upon the profile, threat, environment, routes, facility being visited, etc. I will discuss several, but I prefer one car in the front as a SAP and the other in the rear. What this does during movement is allow the SAP to give feedback to the convoy on road conditions etc. before the lead vehicle gets there. This also allows them to block off intersections or traffic circles so the convoy can get through much faster. And one of the most important uses in my opinion is to have the SAP drive up on overpasses as the convoy approaches and passes, then get back in front. This provides a much greater security bubble on the overpass, since this is a favorite spot for the threat to hit vehicles from. This also allows a counter-attack on two sides of the threat formation. If the convoy is hit after the SAP passes through they will turn around and begin cover fire from the front, while the CAT pulls up and begins cover fire from the rear. If it is a roadblock put in place after the SAP goes by, they can turn around and hit it from behind. Because these techniques and variations can be used either in high or low profile while moving, the SAP/CAT can be very effective and a simple means of increasing your security profile.

At embus/debus locations you can have the SAP block the street to the front of the drop off point and the CAT block the street to the rear as the convoy assumes its position to load or unload the clients. This gives the security detail more control of the surrounding environment. When operating in the low profile mode at these venues the SAP and CAT can assume positions on both sides of the venue and react to any incident from behind as the attention will be on the embus/debus location and the security detail.

STATIC SITE OPERATIONS

In static or fixed site operations at locations such as the client's residence area, office, work sites, and temporary sites there are still going to be threats against your clients depending on the operational environment. Having a CAT or quick response force available to help repel or counterattack any assaults on the site will enhance your security for those sites.

Fixed site permanent areas such as residence and office areas represent more of a challenge for the threat since he is not picking the best place for him to attack. But he has more preparation time, can gather much more intelligence through surveillance and can pick the time and location to conduct his attack, so he still has a great many advantages over the target. The problem with static sites is people become complacent when no incidents occur. Security staff get bored and fall into familiar routines or patterns that help them function in their jobs. Operational security of the clients also suffers and information will start getting out that shouldn't. A CAT or a static site QUICK REACTION FORCE (QRF) will help alleviate some of these problems and enhance your level of protection.

The QRF is best suited for the static role in day-to-day operations around facilities in high threat areas. When deployed by the static security manager, they move to the site of the attack and render assistance by plugging any breach in the perimeter or by moving outside the perimeter and flanking the attacking force.

For use at work sites and locations I had a QRF stationed in vehicles at the main gate, with the engine running, so the driver could plug the gate if someone tried to drive through and the rest of the QRF could help the gate security repel the attackers at their point of entry. Since they had a vehicle they could also respond to other locations on the work site as directed. In another scenario, they could leave the work site and render assistance to any client convoys that had been attacked on the way to the work site. A QRF could also be used to form an extra layer of security or another CORDON through which any threat would have to pass to get to the clients, which would give the static site guards more of an opportunity to move the clients into their designated secure locations.

They can help SEARCH the area for attackers/ intruders after the clients are in their safe areas and under the protection of the static guard force. This will require some knowledge of CQC and a search plan already in place with one person directing the efforts of the searching forces. The most important thing to remember about a quick reaction force is not to deploy them too quickly. Always identify the main attack point, and make sure it is not a ruse to draw security from another sector so they can launch the main assault there.

**NOTES**

**NOTES**

CHAPTER 21
COUNTER-SNIPER OPERATIONS

Most personal security details will not have a reason to deploy counter sniper teams during a mission, but there are several occasions when they should be considered, mostly dealing with public appearances that have been publicized in advance. This includes:

- High profile weddings where it is known the client will attend.

- Ceremonies such as opening a plant or road or receiving the keys to the city or other awards.

- Highly publicized arrivals or departures at known locations, such as airports or train stations.

- When the client will be traveling along a known route at a known time such as a parade.

- When there is a threat of sniper attack based upon your threat management and risk assessments.

Also remember the threat does not have to be specifically targeting your client, they could just be targeting the EVENT for publicity because of its location or because of who is sponsoring it.

PRINCIPLES OF USE

- All snipers on the counter sniper team need to have their OWN ZEROED WEAPON.

- They should do advance recon of the area so they can pick out their best observation and fire points.

- They must deploy in a low profile manner so no one knows they are in the area.

- They must have communications with the security detail at all times.

- The counter-snipers should be able to engage any clearly identified hostile targets or snipers as necessary, communicating when they are shooting and where the target is located.

- All members on the counter sniper detail must have regular, realistic training, shooting from different elevations, angles and platforms to simulate the operational environment.

- THE MEMBERS ON THE COUNTER-SNIPER DETAIL NEED TO BE TRAINED AND EXPERIENCED SNIPERS!!!!!!!!!!!!

COUNTER SNIPER EQUIPMENT:

The sniper on the counter sniper detail needs certain equipment to perform his job. At an absolute minimum he needs:

- A sniper rifle which is capable of 1 inch groups at 100 yards.

- The rifle scope should be a 4-10 variable scope; 4x is the best for precision shooting while 10x will help confirm the hostility of a threat and identity of the proposed target.

- Precision ammunition to allow the sniper to use his weapon to its best capabilities.

- A hands free capabile radio that can communicate with everyone he needs to communicate with.

- Spotting scopes or binoculars for initial observation of his designated area.

Other equipment he can have to improve his performance is:

- Uniform or clothes that will fit into his area of operation

- Ghillie suit, either urban or rural, to help him blend into areas without natural concealment

- Cold or wet weather clothing to negate the effects of weather

- Hard case for transporting his rifle and scope

- Necessary cleaning equipment

- Compass

- Spare batteries

- Sand sock, bean bag, tripod etc.

- Light source

- Camel back for hydration

- Tape recorder for accurate hands free observation remarks

- Digital camera, video recorder

- Some type of covering, camouflage net or canvas

- Food that does not need to be cooked or prepared and water

- Log book and range cards

- Night vision devices

- Photos and maps of the area

- Visual marking devices such as VS17 panels, strobes, etc.

These suggestions are based on commonly used equipment. Each counter-sniper team will have their own requirements based upon their training and experience.

THE COUNTER-SNIPER TEAM

The CS team normally has two people, the primary shooter or sniper and the observer who is also sniper qualified. The tasks for each can be broken down into the following areas.

The primary shooter or sniper:

- Deny the use of all possible sniper positions to the threat.

- Observe and report on movement on roof tops, windows and elevated positions in his area of observation.

- Observe and report an suspicious ground movement, persons and large gatherings.

- Engage any hostile persons/attackers that the PSD cannot engage due to distance or cover.

- Engage and stop any attacking vehicles.

The observer and assistant sniper:

- Assist with setting up position and equipment.

- Handle all communications.

- Assist with windage and ranging, adjust shots for CS as necessary.

- Provide security for the CS.

- Additional fire power as required.

- Help engage multiple targets with precision fire when the CS cannot.

THE FIRING POSITION

First the CS team must go to the mission site to find the areas of vulnerability to the client. They will stand in these areas (embus/debus point, etc.) and identify all possible fields of fire and firing positions for a sniper. Using this information they will then look for a point that allows them to cover all these possible firing positions. More than one CS team might be needed. If a sniper cannot cover all possible firing positions on the areas of vulnerability then the areas of vulnerability should be changed if possible. If not, then access to the possible firing areas should be restricted and those areas searched and secured during the client's time of vulnerability.

When choosing their firing/observation position the CS team not only considers it for its ability to cover all possible threat sniper positions but they also need to consider the following:

- **Range** should not be more than 200 yards at the maximum whenever possible.

- The **field of view** from the CS position must cover all the points that have been designated as possible threat sniper positions and routes of movement for the client.

- The position should have a **flat angle of shot**, if not try to keep it to a minimum.

- **Comfort** of the CS team is important depending on the length of time they need to stay in their position, so make sure you take all possible weather conditions and physical needs into consideration.

- **Light at the position**. During the time of client movement, will it be in your eyes or limit your visibility? This must be checked.

• **Position security**, based on not only the threat, but from local people also. After all you don't want kids or others walking into your position giving it away or inhibiting your shot.

• Don't forget the **media** because they will be trying to get the best positions also for camera shots.

On the day of the mission the CS team needs to be in place early, before mission start for the PSD. They must make range cards, windage estimates, and range estimates and they need to begin observation of the area for suspicious activity, people and vehicles.

CONTROL OF THE CS TEAM

Whether it is one or many counter-sniper teams they all report to the security operations center set up by the advance or whatever is being used to control the security details mission. They should be on the same frequency as the rest of the security detail for the event so they have a better understanding of what it taking place with the client and movement as they are observing the crowd. All CS teams need to be in contact with each other to coordinate efforts while in position.

There also have to be some precautions taken so that the CS teams can identify any other security elements that will be helping supplement their security detail, such as local law enforcement or undercover officers. Usually a pin, such as the type talked about earlier, is worn on the lapel of a coat or uniform so that these people can be picked out quickly by both the security detail and counter-sniper support. The same should be done for all vehicles being used by the different supporting elements.

REPORTING SOP FOR THE COUNTER-SNIPER TEAM

A standard operating procedure and reporting format should be developed by the private security company for all of its sniper and counter-sniper operations. This is important so the reporting system can be learned by everyone and understood by everyone instead of each team or detail having its own. (This usually happens when people have attended different schools and training.) Your system should be as simple as possible since the security operations center, the mission detail and shift leaders and the counter-sniper team will have many other things to worry about during the mission. It needs to be clearly understood, accurate and very concise to keep radio transmissions as short as possible.

LOCATION REPORTING

When a sniper team is calling in a location it is doing so to get other people in the security element to look at that specific location. The first step is identifying a standard building, basically which one are you looking at. Each building should be given a name or designation (with address) during the mission planning. Then you have the building itself. Use a color code system to identify the sides of the building, for example white can be the front, black the rear, red left side, and blue right side. Make sure you always use the same colors for this. Why use colors? Because many people have problems with cardinal directions unless they have a compass out, and when dealing with outside supporting agencies, especially locals, they might not have a clue at all, but when you tell someone white is always the front of the building they grasp that much quicker.

Next will be need to report which doors and windows in a building we want others to look at or to let them know which ones the CS team is looking at.

The simplest way is to work left to right using a two number system the first number being the floor the opening is on, and the second number being its location in line with other openings on that floor, so 2 / 3 would be the second floor, third opening from the left. So now when a CS team calls in a position or someone is calling one in for the CS team they can say "building two, white, 3 / 1" and everyone should know which building, and that they are talking about the front of it, third floor, first opening from the left.

Simple and easily understood when you have lots of radio traffic and other things taking place during an operation.

REPORTING INDIVIDUALS

When reporting individuals to the ground elements, the CS team must be able to describe them so they can be found easily. Keep this system as simple as possible. The easiest way to identify a person is by clothes or other characteristics that stand out. A simple system is using the ABCs these are:

• **AGE**: Approximate age of the subject

• **BUILD**: There are only 3 that need to be used - slim, fat or athletic

• **CLOTHING**: What the person is wearing that is easily identifiable (telling them what color shoes someone is wearing is going to be hard to see in a crowd, give them something from the waist up so it can be seen)

• **DISTINGUISHING MARKS**: Marks, scars, tattoos

• **FACE**: Type of face - long, round, double chin

• **GAIT**: How the subject walks - fast, slow, limps, swaggers, struts

• **HAIR**: Color of hair, length, how it is cut

• **SEX**: Male or female

You can add any other information that will help a subject be picked out more easily, such as props they are carrying, like bags, a back pack, a camera, umbrella, shopping bag or the direction the suspect is traveling and location such as near the road or close to the building. I don't usually give a height except for short or tall, because this is so subjective that two people don't usually agree with what 5'9 is at a distance. The CS team might be perfect at this but the guys on the ground may suck.

When directing someone to the subject use the clock method from the point of view of the person you are directing, not from your position. This will allow the person you are directing to orient faster and get to the subject in a timelier manner.

REPORTING VEHICLES

This needs to be kept as simple as possible also, keeping the format and the description easy for everyone to recognize:

• **SHAPE**: The shape of the vehicle - 4 door, 2 door, van, sedan, hatch back, SUV

• **COLOR**: Basic color of the vehicle - red, blue, white, gray, etc.

• **LICENSE PLATE**: If you can see it and the people you are directing towards it can see it, this will direct them straight to the correct vehicle.

• **IDENTIFYING MARKS**: Something that stands out - bicycle rack on top, antennas, bumper stickers, body damage, window damage

• **NUMBER OF OCCUPANTS**: If possible give the number of people you can positively see. If you cannot tell don't say anything because a bad call on this could cause someone to pass up the correct vehicle.

You also need to give the direction the vehicle is heading, lane of traffic it is in and speed it is traveling. All these will help the ground security element or the CS team to locate and, if necessary, engage any threat vehicles.

A counter-sniper team adds to any security environment where they are used. While they are not necessary at all times for all missions, when they are needed and used they can provide essential real time intelligence to the detail leader, accurate evaluations of people and vehicles in the mission area and precision fire to neutralize threats that are outside the security sphere provided by the on ground detail. This is a resource that may be needed so it is vital that some members of the security detail be trained in these skill sets and have experience using them. While they might not get a chance to use them at least they are available to the detail leader if there is a need.

CHAPTER 22
BUILDING SECURITY CONSIDERATIONS

As a member of a personal security company you can do either client transportation or static security at one of the client's primary fixed sites. Protecting the client at a static location is just as important as protecting him while moving. This chapter will give you a quick overview of setting up security at a static location such as a residence, office or a temporary site being visited or other area the clients may need to visit.

There are basically only two types of areas where a site can be located, either an urban site or a rural site, and both have pros and cons.

URBAN sites have a better emergency response from local services than rural, and an urban site also has better communications capability and coverage. You have many more choices of routes and better maintained road surfaces in and out of the site and to other sites so it has easier access. Since you have some different route choice this means you will also have more and easier ways to evacuate if necessary. There is a lack of privacy since there will be much more vehicle and pedestrian traffic, which in turn provides more cover for any threat activity taking place, allowing them to get closer to the site with their surveillance since they are able to blend into the local operating environment. It is possible to have buildings or other structures that overlook the client's area that can be points of observation for the threat or areas where stand off attacks can be launched from.

RURAL sites have more privacy since they have less dense pedestrian or vehicle traffic. Because of this much population density it is much more difficult for the threat to conduct surveillance in general because it will be harder to blend with the locals and almost impossible to conduct any close surveillance. Do not, however, forget that the locals in any location could be paid or coerced to do this for the threat.

Local emergency services will have a much slower response time because of location or distance and might not be as well equipped. The roads won't be as well maintained leading to much more limited routes to and from the site, which will make you more predictable and easier to watch or attack. This in turn will make your choice of possible evacuation routes limited. Communications will be more difficult depending on the type used. Cell phone coverage will be worse in this area unless a tower is near so cell phones could be hit and miss without a signal booster. Repair and maintenance of communications lines will not be done as quickly as in an urban environment.

You will more than likely have much more physical ground to cover and protect since houses in the country tend to have greater areas of yard and the lay of the land around the facility may have areas where you lack direct observation. There also may be high ground around the location that will need to observed or controlled.

With any structure on the grounds of the main facility or building you have external considerations and internal considerations for security purposes.

EXTERNAL considerations include:

1. The **location** of the building/facility which is a major factor when considering security.

- Is it by a high speed avenue of approach?

- Is it located by any main thoroughfares?

- Is it surrounded by an outer wall or fence or something that blocks it off from these?

- What is in the vicinity around the structure, landscaping, woods, shrubs, cell towers, buildings, high rises etc?

All these things need to be considered and steps taken to ensure these things will not give an advantage to the threat and that you have some modicum of control over them.

2. **Perimeter.**

- Is there a physical barrier in place? What type?

- Does it give the degree of protection required for the threat environment?

- How many breaks or entrances are in it?

- Are they controlled in some way?

- Is the fence lighted sufficiently?

- Is the barrier alarmed? What type? Where is it monitored?

3. **The yard**.

- What is physically on the grounds?

- Are there trees, shrubs, and other vegetation that not only screens the facility from outside view but screens the barrier from security?

- Is there any high ground or areas of limited or no visibility?

- Are there things like a small out structures, vegetation, or smaller walls that will cover anyone from the view of the security force?

4. The **buildings** on the grounds including the main office or residence.

- What type are they?

- How many?

- How often are they used or checked?

- Are there existing security systems in place in these buildings?

- How often are they checked?

- Are the doors alarmed? Blast resistant?

- Are the windows and skylights alarmed?

- Do they have bars or grills?

- Can these be opened from the inside?

- Are they bullet and blast resistant?

5. The **driveway** should have barriers that people have to drive around and speed bumps. Also all parking places for visitors and/or deliveries should be away from the facility.

INTERNAL considerations:

1. The **main suite or rooms** the clients live in should be self contained for things like rest rooms, have telephones and other communications hook ups as necessary and have doors with locks and a panic button.

2. A **safe room** should be constructed in any building where the client or clients spend a lot of time. The safe room will have fortified walls, doors, and windows. It will have secondary means of communication, a supply of food and water, medical supplies, weapons and body armor, and fire extinguishers. This room is a place that the clients can go to or be taken to by security if there is an incident at that location. They will stay in the room until a known element comes to evacuate them.

3. A **security operations room** needs to be located in the facility. It should be staffed 24/7, have land lines dedicated to it, cell phones and radio communications that can reach emergency services, weapons, emergency equipment, and medical packs. This room is always locked and only authorized personnel are allowed in or should know the location, which means it is cleaned and maintained by the security staff.

4. There should be rooms set aside for **staff and security personnel** who work on the site, so any outside surveillance will not know the number of security personnel on site at any one time.

5. All **guest rooms** should be self contained with a private bath and be as far away from the clients and security as possible. The corridors outside guest rooms should be monitored and security placed to ensure guests do not wander into parts of the facility or building where they have no business.

6. Any **public rooms** need to have some type of monitoring in place, and doors leading to parts of the house not open to nonresidents need to be blocked off or have controlled access.

7. All **public areas** such as gardens and parking areas need to be monitored for who comes in and who leaves. There should also be temporary fencing or barriers placed up to make sure that nonresidents do not have access to areas on the grounds where they should not be.

OTHER CONSIDERATIONS

TELEPHONES: Land lines should be installed away from windows or any other openings where someone outside can see in. Have an SOP in place in case the telephones stop working, have emergency numbers on speed dial on all phones and have caller ID on all phones if available.

TRASH: All trash from the residence should be burned. At a minimum all papers should be shredded and burned not thrown away. Random checks should be done of the trash awaiting pick up to ensure no company, operational or personal information is thrown out.

Security done at any facility or location where the client will be spending time, whether it is a residence, barracks type area, work site, office, or temporary site at a work start up location should all be treated basically the same. They need a layered security plan that consists of dividing the security into areas of coverage: the outer layer and the inner layer.

OUTER LAYER SECURITY

The outer layer of coverage consists of the area from the outermost perimeter that is under the control of security up to the walls of the facility including the perimeter fence, access control and some type of physical barrier to channel movement to controlled access points. **ALARMS** need to perform 3 functions:

- Detection of unauthorized people

- Identification of the location of intrusion

- Response of security or local law enforcement and internal security to protect the client

If an alarm sounds, portions of the security element should go to the client's rooms and provide protection in place or move them to a safe area or safe room. At the same time, other designated personnel will respond to the alarm site. There has to be sufficient time for this to happen AFTER an alarm goes off regardless of location.

INNER LAYER SECURITY

The inner layer of coverage is everything inside the walls. Inside the walls of the building the security detail needs to maintain good access control, establish a badging system that will identify who can go where and a way to restrict access to areas where people do not belong or have business. The security center will monitor all activity at the facility. They should answer all the phones coming into the facility screening calls and routing them safely. Most will say that this is a secretary's job but lots of information gets passed inadvertently through bad operational security by the untrained.

The security operations center will maintain a log, round sheets, and if any video and audio surveillance is used, it needs to be maintained for a set period, so it can be reviewed. Usually the security detail for any facility is permanent, so they know the grounds, personnel and daily, weekly and monthly routines and operations of the site. Sometimes you can rotate security personnel between doing personal security details and fixed site security as long as the people rotating know the jobs and not everyone is rotated out or in at one time.

DELIVERIES

To keep the security level at its highest, you should also standardize deliveries to the work site or residence. Check out what is delivered, and if possible call the company and get descriptions and names of personnel who are making scheduled deliveries and the type vehicle they are using. Search everything. For unscheduled deliveries find out the reason, who authorized it and confirm it with that person and search everything.

CONDUCT

Conduct of any security detail that is around the living area or work area of the client is important. Remember this is an area where people live, and in some places it could be lots of people with different religions and ethnic backgrounds. Be aware of what is around you and act accordingly, especially around the work areas when you are off duty. Being off duty does not mean you can act however you want or do anything you want. The clients have still have meetings and work visits. NOTE: I took some clients to meet with work partners. We got received and escorted upstairs to a meeting room, which overlooked the living area, where there was a volleyball court and the off duty members of their detail were out there tanning in the nude. Two of the people transported were women. One was a local, and the detail's behavior didn't go over very well with our local nationals that were there.

You should not smoke while on duty because it distracts from doing your job, takes away your ability to smell, and at night gives away your position. Always keep a low profile and try to become part of the background that everyone knows is there but few notice. Always practice weapons safety, especially in the static site, so no accidental discharges or dropping weapons. If you look like the keystone cops you will lose some respect of your clients and might cause them to lose respect of their clients or business partners.

VISITORS

As part of static security someone needs to meet visitors and make sure they know where to go. They might have items such as body armor that need to be stored until they get ready to leave. Introduce yourself and let them know they can contact any security person if they have problems. You should be tactful when working with your client and his clients.

RANDOM SECURITY CHECKS

Site security, whether inside or out, should conduct scheduled and random security checks throughout the entire day. This will make you unpredictable and show a strong security presence to anyone conducting surveillance. The staff of the site should be monitored also. Most clients will hire local nationals to do daily work like cleaning and maintenance and all portions of this should be monitored and scheduled so that the staff can be observed. This is not only to protect the client from his local hires but in some cases to protect the local hires from abuse or sexual harassment. When cleaning rooms the staff should leave the doors and window shades open while they are working, even if a client is present in the room. This will increase their visibility, allow anyone passing by to see into the room, and allow security to observe inside the rooms while on rounds.

If you have female clients or female staff, you should have female security personnel available. This is important when searching local staff when they come in and leave, and in the event of an incident with a female client, leaving a female security guard to watch and monitor a female client will take a lot of tension out of the situation.

When working with a local staff, especially in high threat environments, operational security is very important and can be vital to normal client operations and work but is always vital to security. Do not allow any local nationals or visitors into areas that have not been sanitized of client work information, and never allow anyone but security into the security operations center. Its not that big of a deal to sweep and clean your own bathroom.

OTHER SECURITY DETAILS

As a static security manager you will probably be receiving other clients or people who work with your client and THEIR SECURITY DETAILS. Treat them with professional respect, the same way you want to be treated. They are doing a job also. If they want a BG or detail leader to accompany them to a meeting, don't raise a fuss. Have an area where security details can park their vehicles and an area set aside for them to get out of their gear and relax; I set aside a trailer next to the secure parking area. It had a working bathroom, TV, water, soda, etc. so visiting details could have somewhere to be besides standing out in the sun or sitting in their vehicles waiting for hours. Talk to them about the route they used and conditions to find info to add to your database for your future reference.

NIGHT SHIFT

How you work at night is different than how you work during the day because there should be less pedestrian and vehicle traffic into and out of the facility. You still conduct security checks, but you can seal those access points that will not be in use at night and set alarms on them.

Motion lights are something I like to use along with regular lights so that it can be known when something or someone is passing in that area and if it is not a security detail member it can be checked out. Make sure all lights, cameras and alarms are working properly. Security personnel should work with radio ear pieces in, and should not talk unless necessary to avoid giving away their positions and any information that can be gained from listening.

For the inner perimeter the same things apply. Put motion lights in corridors and rooms that do not have a presence at night, to show you when someone is walking down that corridor or is in that room. If you leave the lights on all the time it does not stop people from using these areas and you might not notice people who are in there. You should always close all curtains and blinds before turning on interior room lights so anyone watching the facility will not be able to get a layout of the room or interior of the structure.

Once a door or gate is sealed the alarm should be set and the seal remain unbroken until morning, then in the morning each seal number is checked against the log to make sure someone didn't use that access point at night and reseal it.

FIRE SAFETY

Fire safety is another thing that the static site security will have to work with and be responsible for. Security should know the location of all fire fighting equipment and should ensure it is serviceable. Drills should be run both day and night shift for maintaining security while fire fighting. The detail leader will decide when an outside emergency service is allowed on the site. Don't automatically let in emergency vehicles. If you must evacuate the site, then evacuate and account for all people, and move the clients to another safe area.

INDIRECT FIRE

There should be SOPs in place for attacks, or ground assaults against a site and these should be rehearsed. At some work locations, bunkers may have to be built at different areas on site so that clients can get to the bunkers regardless of where they are working. These bunkers should have rations, water, body armor with helmets and some type of communication so that security will know where each person is. There should be a designated all clear signal to let everyone know it is safe to come out, then a damage assessment should be done.

The same is true of residence areas. Residence bunkers should be spaced evenly around the living area, and should be stocked with the same supplies but they should add a flashlight or chemical lights for night use. Security will have to check room by room in a residence area to account for everyone. This is why it is important for people to sign in and out at the access points, so security doesn't waste time looking for someone who is not there. Once the all clear is given, security should do a search of the area to look for impact points and any duds, and then the appropriate EOD should be called to take care of it. Each person at the work place or residence should know the bunker they need to go to.

Having assigned bunkers will greatly enhance the ability of security to account for everyone during and after an incident. Security should have assigned areas to report to, some to the perimeter and some to bunkers to be bunker captains and account for people who are supposed to be there and keep everyone calm. Don't forget about your day work staff. Day hires and local nationals will need areas to go to also. So when you are planning your bunker layout don't forget to account for them in your personnel count.

BOMB THREATS

When considering a bomb threat, first consider how it was brought in. By phone call? Was a note left? Then you need to consider the validity of the threat. If it is at a static site with 24 hour security provided by your company, what are the odds that a bomb made it onto the premises? If it is at an away site where security is someone else's responsibility, inform the site security and move your client to a safe area. Or if you feel the threat is great enough evacuate him to another location.

Get as much information as possible if you are receiving the threat by phone. If it is a credible threat evacuate everyone out of the buildings and into

their bunkers, and bring in EOD. If it is at an away site evacuate the clients but consider the possibility it could be an attempt to get your client out of a secure location to make him an easier target during movement.

Part of your security plan should be scheduled and random sweeps of static sites with EOD teams and dogs. I would also use dogs at all the access points. NOTE: Working dogs can only work a certain amount of time before they need rest. I would use working dogs at random but keep other dogs at access points in view of the general public with signs notifying everyone that they will be searched by use of a dog when entering. These dogs do not have to be working dogs. I have used strays that we fed, got shots for and kept housed at each access point. The perception of having the dogs there will be noticed by any surveillance on the location. These non-trained dogs need to be treated exactly the same as the true EOD dogs to act as a visible deterrence to the threat.

Static site security is not as exciting as conducting PSD mobile operations so it can be very boring, but it is one of the biggest parts of any security operation since the client will need safe areas to live and work. This also gives a secure location for the client to move from and return to and a place that all personnel, including the security details, can relax, feel safe and get that much needed rest to do their mission well.

TEMPORARY SITE SECURITY

On some missions the PSD will probably have to provide temporary site security at the facilities or locations being visited, especially at work sites or proposed work sites, warehouses, etc. While local security will probably be there, usually it is very low priority untrained personnel with few or no weapons or ammunition and no physical security at all other than opening the entrance gate to the grounds. As the detail leader you will still need as much coverage of the area as possible. 360 degrees is the perfect answer but not always achievable, especially with the resources available on a movement.

Prioritize your security in your security sphere. As you pull into a site, look over the site layout and see where your client needs to be for his mission requirement, and then work out from there. First you secure his mission area. Leaving the client or clients in the vehicles, you send in someone to make initial contact and search the building and meeting area while two other detail members search the outside of the structure, looking for all exits and entrances. Once these are found, they will be secured to the best of your ability with the resources you have.

In a 3 vehicle mission convoy place the lead or scout vehicle on one corner so the driver can observe two sides of the building. The main body will be on the opposite side so that driver can observe the other two sides. The tail vehicle or gun ship will be in front of the main gate to the facility with its rear facing the gate so your heavy firepower is covering your weakest entry point. The rear gunner and the driver stay with the tail vehicle. The other three detail members will be in the structure the client is in with one at the main entrance, one in the hallway where the client and detail leader are located, and another by the door to the meeting room. The driver and shift leader will remain with the lead vehicle pulling security on two sides. The other two detail members will be on the roof of the building where the client is located with a sniper weapon and spotting scope in addition to their normal weapons. They will be watching for any unusual movement outside the perimeter and inside. The main body will just have the driver who will be watching his area on the outside of the structure.

Once all is in place the clients go in with the detail leader. The clients will keep their protective gear with them. If they are not wearing it, then it will be in the same room with them at all times. This is a very simple plan that provides the maximum coverage with the people available. The larger your detail the greater security you can provide. Good mission planning and threat management will let you know your requirements so you can have a tentative plan in place before you arrive.

There are other things you can travel with to help set up temporary security at any site, including empty sandbags, a shovel to fill them, concertina wire which can be carried on your vehicle along with fence pickets to hold it in place, long heavy duty screws with a cordless drill to seal doors and windows, duct tape, cloth or paint to cover windows, chains with locks to secure gates and canvas or umbrellas to provide shade for those people operating on the roof or outside the vehicles since some areas will get extremely hot for people standing in the sun for several hours. These items do not take up much space and can be invaluable when securing a site on a temporary basis.

APPENDIX
TYPES OF SURVEYS

There are four types of Surveys that are normally conducted by an advance, these are: General Site Survey, Remain Overnight Site Survey (RON), Airport Survey and Hospital Survey.

GENERAL SITE SURVEY

The General Site Survey is utilized for a general site, such as an office building, museum, baseball park, restaurant, store, cemetery or outside event. It is not site specific and is flexible enough to be adapted to most general areas you encounter.

REMAIN OVERNIGHT SITE SURVEY (RON)

Remain overnight (RON) surveys are used when the client plans to stay overnight at a location such as a hotel, guesthouse, ambassador residence or other facility.

AIRPORT SURVEY

Like the name implies this survey is site specific for an airport or air terminal location that the client will be flying from or flying into. This survey can be modified easily and used for rail terminals if the client is arriving or departing from a rail line.

HOSPITAL SURVEY

This type of survey is site specific to a hospital, clinic or other medical facility that the client my have to visit or for one that is a safe haven along a travel route.

GENERAL SITE SURVEY

1. **Identify location of mission**: Identify the location for the Site Survey where the client will be visiting. This information can be obtained from the Itinerary.

2. **Identify the required POCs**: Usually the person hosting the event, his security manager or another member of his staff will be the primary POC. If you cannot locate a POC before deploying for the advance, contact the client or client company and ask for the name and number of the person they have been dealing with in setting up the event. You must ensure you have a way to make contact with those people identified as your POCs. It will also be necessary to provide the POC a contact number for you and the security operations center in case they cannot get in touch with you and there are changes to anything you have already discussed.

3. **Motorcade arrival and departure area** (Primary and Alternate): At the mission site you will need to coordinate a primary and alternate arrival and departure area for the convoy. In choosing these areas you need to keep in mind the likes and dislikes of your client, crowds and location of press personnel and the size of the vehicles being used. Some clients will not want to be dropped off in the immediate view of the public. Others prefer discreet areas where the clients can debus without being noticed. Another thing to think about is you must not block any handicapped access areas or parking areas and stay out of emergency lanes; this could cause an embarrassing situation.

4. **Convoy vehicle parking and staging**: Coordinate parking and staging for all the vehicles that will be traveling in the security detail convoy. Whenever possible, get secure parking areas, or one that has restricted and/or controlled access. You need to keep in mind the location of the parking area as it relates to the location of the arrival and departure area. In some cases the parking and staging areas may be too far away from the arrival and departure area. When

this happens the shift leader will need to give advance warning in order to pick up the client on time for his scheduled departure. Try to keep the convoy vehicles as near the arrival and departure area as possible. Always have one vehicle manned and stationed so that an emergency pick up of the client can take place in case of an incident.

5. **Actions upon arrival/escort for client**: Upon arrival, advance personnel will be standing at curbside to direct the convoy vehicles where to stop for drop off. The advance leader will escort the host or the host POC to where they will conduct an introduction to the client. Another member of the advance will inform the drivers of the convoy vehicles where to park. If there is more than one client getting out of the main body they will all exit out of one door, which will be the door closest to the entrance of the facility.

6. Any **walking routes to an event or function** (primary and alternate) should be obtained beforehand and a recon conducted then annotated on the site survey form and a strip map should be provided to the detail leader and or bodyguard.

7. **Identify private holding area(s) and safe havens** at the site and or sites. Safe areas should be located at each site for the client or clients. A holding area is a protected place where the client can wait until he is to appear at the event and where the client can have some privacy because some clients will rehearse speeches in these areas. This is also the place where they can prepare for the event in private such as having a microphone placed on them or a place where pictures are taken with special visitors or people.

The holding area should have been checked for any electronic surveillance so it is a place where private phone calls can be made and where one-on-one business conversations can take place away from a crowd. These areas should be large enough to hold the client or clients and should be equipped to handle any foreseen needs such as rest rooms, refreshments, chairs, phone, internet, fax machine, etc.

Safe havens should be located throughout any site that a client is visiting regardless of the duration of the visit. A safe haven is an area with limited access where the client or clients can be placed and defended by security until the incident is handled by law enforcement or site security, or evacuation can be done safely. Directions to these areas need to be annotated on site survey forms, along with any key codes or other requirements to access them.

8. Identify **evacuation routes** at each site. You should have both primary and alternate routes. If the site or facility is large and the client will be visiting different levels and venues then there may be a need for additional evacuation routes for each location visited within the facility.

9. Location of secure **land line telephone, computer terminal, fax machine** and instructions on how to use them. The advance will need to confirm with the client the possible need of communications equipment during the mission for personal or business use. It is important that the equipment be secured to help safeguard business information, contract negotiations and/or trade secrets.

A land line telephone may be needed if cell phones will not function inside the facility or site, so it will be necessary for the advance to test this cell phone coverage. Communications equipment should be located in the holding area or room that is set aside for the client's use. The advance will obtain the telephone number for that phone and fax machine along with the necessary dialing instructions. The advance will pass this information to the security operations center so that both lines can be tested before the client arrives to make sure they function correctly and their use is understood.

You might need several telephone lines if cell phone use is not available inside the facility. The other phones may be needed for other clients or security

detail personnel for coordination. An internet connection might also be required by the client, so make sure one is available in the holding room if necessary.

10. The advance needs to obtain the location of **bathroom facilities** located throughout the site that is being visited. They will need to know the location of both male and female facilities in case one of the clients is female. It is preferable to gain access to facilities that are private since they are usually better maintained and have better controlled access than public use facilities.

11. Location of **elevators/escalators and stairwells** at the site. The advance will need to know the specifications for the elevators at the site, including obtaining the maximum weight to be carried and maximum occupants each can hold. The advance should try to coordinate the locking down of an elevator for the client when he arrives. This may not be possible due to needs of the host or site security people so the advance could just station one of his people at the elevator to hold it and keep others from entering when the client arrives at the debus location.

The advance should test the elevators, ensure they are in good working order and that they know the movement capabilities of the elevators. What I mean by this is some elevators only go to certain floors such as floor 2-10, and to go higher you have to use a different elevator. Know the client's destinations at the facilities and make sure you know which elevator will get him there.

The advance should also become familiar with the different types of elevators and floor numbering system for that facility, because sometimes floor numbers listed for elevators in foreign countries may be different than what you are familiar with. A good example of this is in Europe and European settled countries, floor one is the ground floor, but in some places the ground floor is designated as M for main or L for lobby. If there is going to be lots of equipment or luggage traveling with the mission, the advance might want to ensure there is a freight elevator available. If the event at the site calls for formal wear and you have females who are attending in evening dresses you may want to ensure that escalators are turned off, so that high heels and gowns will not get caught. Also you may want to check how slick the floors are since most hard soled dress shoes will slide easily on a slick surface, so you can warn the client and security detail members if necessary.

12. Identify **security capabilities at the mission location** and in the surrounding buildings. It is important to identify the security capabilities of the site that will be the main area of the visit and the surrounding buildings and facilities. Does the mission site have its own security staff? What is their training/experience? What are their hours of operation? What are their rules of engagement for incidents and for emergencies? Are they armed? Do they have facility wide communications? Do they have a presence outside the site or do they only maintain security inside? Is there a presence of local law enforcement, either on duty or off duty? Do the site security personnel work in uniforms, plainclothes or a combination of both?

You will also need to find out if they have a security operations center and if they have surveillance cameras and the areas of coverage for the cameras inside and out. Do they record them and how long do they maintain the recordings? Do they provide any coverage for parking areas? For facilities around the mission site, these same types of questions need to be asked, especially about security, and surveillance detection capabilities and surveillance cameras outside. It would do no good to secure the mission site only to have an attack initiated in the building across the street or from next door.

After the event, it's also good to get copies of any surveillance videos from surrounding buildings and, if possible, from the mission site for review and to look for any surveillance that was taking place against your security detail and the client.

13. **Access Requirements** for the mission site, venues and surrounding area. Sites that already have their own security program may have access requirements where badges or permits are required for any visitors including the advance, the security detail and the client or clients. The advance will need to coordinate with the event and mission site security staff to get all the necessary paperwork required to meet the access requirements and they should obtain passes, vehicle passes, pins, badges, whatever will be required before the detail when the client shows up. This will help the operation run smoothly and cut down exposure time when moving vehicles and the client into the secure restricted access areas.

14. **Obtain street maps, city maps, floor plans and schematics** for the mission site/facility, surrounding structures, and parking garages. The advance should always obtain a copy of street maps, city maps, floor plans and schematics of the mission site and adjacent buildings and structures while doing the pre-advance work. This type of information can be used in preplanning, setting up security, planning evacuation routes, safe havens and for other contingencies and/or emergencies.

15. What are the **media considerations**? In some places and at some events the media are going to be necessary for publicity, such as when a company opens up a school or power plant in Iraq. The advance should always ask the client and/or the host if there is going to be press or media at the site at any time, who notified them, what information they were given and what they are allowed to do. When you get this information, if it was not provided by the client, then you need to contact him and let him know.

Also the security detail leader must know, so he can adjust his security posture during movement since the information is now public. You also need to know if the client is going to want to speak to the press or if they want to avoid it, and you should determine with host site security if they are going to coordinate a "press only area" to control the press personnel, or allow them the run of the event. Also find out what badges and or passes the press are going to be issued so you can inform the security detail.

16. The advanced needs to find out if the host of the event is planning to have a **gift for the client**. In some parts of the world it is a part of the culture to present small gifts. If this is the plan, the advance needs to find out what will be given to the client and what will be a suitable item for the client to give to the gift presenter. It would be embarrassing to the client if he was given a host gift and had nothing to give in return, and it could also hurt the rapport the client or the client company has built up with this individual if he is someone of importance like a local mayor or local clan/tribal chief. It probably wouldn't hurt to bring several different gifts and keep them staged at the mission site so they can be presented if you are unable to find out about the gift exchange in advance.

EXAMPLE FACILITY CHECKLIST

1. IDENTIFICATION OF THE FACILITY by name and mission number:

a. **Local and official name**: Include all known names for the mission site - official names and designations and unofficial names, nicknames, and designations by which the mission site or venue is referred to. (If you know the reason for any nicknames or unofficial names, make a note.)

b. **Functional of mission site**: What is the primary function of the building? Who owns it? What, if any, are the secondary functions of the mission site? (i.e. Cell tower on roof, major restaurant on ground floor open to everyone, bank on premises, etc.)

c. **General location**: Give a physical location of the mission site and where it sits in the city by district and or neighborhood. Also give its relation to any major road network and streets, public transportation nodes around the mission site and any well known reference points.

d. **Local address**: The mailing address and any local variants.

e. **Map references**: If available list all the maps used. If official DOD maps are used, list the title, sheet data, etc. If locally purchased or satellite maps are used, list as much info about the maps as possible. If you use a computer mapping program like Street Atlas, make sure you have the program version number and include copies of the maps with the survey.

f. **Coordinates**: Provide a complete geographic coordinate and any local variant (many maps in Iraq were Russian in origin, using a little different coordinate system). Reference which map was used to obtain/plot the coordinates you are using. When using a GPS, include the type, how many satellites you got, date, time and datum used and your location when the reading was taken.

g. You must provide a **magnetic azimuth and distance** from at least of three significant visible reference points, either natural or manmade, which are easily visible in an urban environment. This will help assist someone in finding the structure on one of the maps, from the air by helicopter or when on the ground.

h. This paragraph will include the names of the people **who prepared the report, date, time, reason and any additional information** that will aid in further identification of the mission site which has not already been covered. This section will also include any problem areas regarding the use of available maps or other references.

2. AREA SURROUNDING MISSION SITE

a. **General Information on the area** surrounding the mission site:

(1) The advance will need a **general description of the area surrounding the mission site**. This description should extend for up to two city blocks on each side of the mission site. It should describe what borders the mission site to include other facilities, streets, key terrain and vegetation, and the locations of any other key structures that may be referred to later. You should include sketches/overlays and photographs of the area described to help support the written narrative and give the rest of the detail or any future details a better grasp of the working environment.

(2) **Additional Information/considerations**: Use this paragraph for any information you think may be of benefit to planning of the mission or any contingency operations.

b. **Security operations considerations**:

(1) **Security centers and control points**: The advance needs to give a description of all security control points found in an area surrounding the mission site. A list of what is available at each one should be provided as well as the following: covered and concealed entrances and exits, water, rest rooms (male and female), electricity, telephone, computer line, and available working space. You will need to have all POCs listed with phone number, area access requirements, the recommended approach routes to and from these areas and any security or other information that is essential for contingency operations. This section will also include sketches/overlays showing location, recommended routes, security locations, and photographs depicting entrances and exits.

(2) Are there any locations that can be considered **Key Terrain** and used for **Sniper/Observer/Security positions** if the working environment calls for it? If so, give a brief description and the location of all recommended sniper/observer/security positions found within the surrounding area. Include distance and angle to the center of the area to be covered from each position and any other data that may affect contingency operations. This section will also include the following information if it applies to the mission and the working environment: a description of each separate location, fields of view and fields of fire, description of how to access the positions, description of cover from small arms fire and concealment from observation, height of the position, who controls the position or access to it, type of glass (if observer has to look through or shoot through glass), does the window open, lighting in area, what type, intensity, when it is turned on/off, if it is automatic, where is the power box, whether it will interfere with observation from that site, any animals such as dogs that could compromise the position, indoors or outdoors (potential effect of weather).

TYPES OF SURVEYS **179**

(3) **Debus or drop-off point for the client**. Each possible debus point needs to be looked at and described in detail. These locations need to be able to support the convoy vehicles, the required number of security personnel as determined by the detail leader, and any special needs that may arise. There should be some ability to control foot and vehicle traffic around the location, and if possible be a covered location or an indoor one. Depending on the site, the site host and the type of event, this location could be determined by the facility security. If this is the case, you will need to work with them and ensure it meets the needs of the client.

(4) Recommend the **route in and the approach for the debus locations**. There should be more than one route to this location so you are not passing through a choke point at the end of your trip to the mission site. If there are other VIPs or guests attending the event and location you should consider using a different area and/or debus location for your client or a different arrival time. If so many important people are using one route or one location it may be too tempting as a target for the threat. Don't forget to designate the route to the vehicle staging area once the vehicles pull away, so to the drivers know where to go to prepare for the movement out of the mission site. Remember, if possible you want concealment until the client gets into the facility. If there is no awning or outdoor overhead, and the environment warrants a sniper threat, the advance should consider using black umbrellas as concealment to escort the clients into the facility.

(5) You need to locate and describe any **choke points** and possible danger areas around the mission site for both vehicle and foot traffic, including locations of any security forces both public and private that may be hostile towards the client, any facilities nearby that may be hostile towards the client such as universities, political headquarters, religious facilities or embassies, the normal traffic conditions in the immediate vicinity, checkpoints, police patrols, construction sites, police or military facilities or installations, public schools, public transportation nodes (rail or bus stations) and exceptionally well or poorly lit areas as well as

alternate routes to circumvent or bypass these choke points.

(6) **Additional Information**: Anything of value not listed above, that you feel will help in mission planning and will help the security detail during movement to the site and for the security mission at the site.

3. AREA DESCRIPTION:

a. A **general description of the area** immediately surrounding the site. In urban areas this will be difficult but it needs to provide an excellent mental picture of the outside area of the mission site without the reader seeing a photograph. Include the general shape, borders, general location of the structure on the grounds, secondary structures, foot paths, vehicle paths, avenues of approach, terrain and vegetation, basically anything will help in developing an accurate mental picture, so when they do look at sketches and photos of the area the security detail will be able to pick up on things they might have overlooked without the written description of the area. It is always best to read the description then go to the photos then the maps to get a good feel of the area before you visit. DO NOT forget to describe it at night or during hours of limited visibility.

b. **Barriers**: Give a description of the barriers around the mission site itself. This description should include what its made of, height, thickness, width, top guard construction, construction of settings of major supports (concrete, etc), diameter of major supports/poles, distance between supports, alarms and sensors, entrances, openings, and weak points. This does not only mean barriers around the outside perimeter, but any other barriers such as fencing on the roof or interior barriers in the parking garage. You also need to include any barriers in the surrounding area, like the buildings around the mission site or other protected facilities close to it such as a warehouse area. Include a narrative of how these barriers will help or hurt the security mission, what actions need to be taken if they are insufficient, along with sketches and photographs.

c. **Perceived areas of weakness**: All the perceived areas of weakness in the mission site security need to be discussed here. You should include each location, the weaknesses, how to correct the weaknesses, and the advantages and disadvantages if any of using it as is without changes. Use sketches/overlays and photographs to help support your observations.

d. **Exterior lighting around mission site**: You need to describe the type, locations, and degree of illumination and source for lighting on the mission site, facilities around the mission site, and for routes leading to the mission site. This description will include how well lit the general areas are, what space is not covered by any lighting, lighting that is not working, and any dead spaces in outside lighting either around the building or around the perimeter. You need to look at and test all backup systems including emergency power sources and activation/shut off sequences and any time delays and the duration of break in coverage. Have sketches and photos depicting light locations and dead space.

e. **Storm sewers, drainage pipes and overflow barriers**: Describe the sewage and drainage system that supports the mission site and the roads that access the mission site. A special emphasis should be placed on any pipes that go under the roadways and walkways or run by entrances to and from the facility and who has access to those pipes. Attention needs to be paid to the different locations to see if, in the event an explosive was placed in them, could they do damage to the client vehicle or to the client? Also, pipe sizes are important to know. Can they hold enough explosive to do any damage or could threat personnel use these areas for a hiding place? Blueprints of the system would be very helpful, especially when conducting the EOD sweeps.

f. **Other Structures inside the mission site perimeter**: You will need to locate and describe any additional structures located in the mission site perimeter or other venues the client will be passing through. This description will include the physical location, dimensions, wall composition, roof composition, floor composition, doors/locks, windows/locks, security, the structure's purpose and who controls access to it.

4. BUILDING DESCRIPTION:

a. A **description of outside of the building**: You will need to give a general description of the building including the basic style, type of material, composition, shape and interior configuration. Photographs will be attached.

b. **Entrances to facility**: Describe each entrance into the structure used for normal day to day access, including the location (name and/or number), composition, direction of opening, locking mechanisms, who controls the keys and master keys, are they monitored, if so how and where, and any other significant data. Each description will be supported by photographs.

Break down the paragraph in the following manner:

c. **Interior description of building**: Refer to blueprints; you should have a general description of the layout of the building by floor or level. This should be short and to the point. Discuss floor plans for each level, including the type of basic construction, floor covering, dimensions if appropriate, and any information that may make this floor stand out or unique. Provide a description of all halls and corridors located in the structure, which will include shape, width, length, height, type of lighting, and floor coverings (if different). You should identify separate corridors by name or number designation for easy identification. Next you will cover all stairways within the structure. Include construction and type coverings on steps, number of stairs per flight, number of landings per stairway, dimensions of steps and landings, lighting, construction and location of banisters, usable width of the stairway, and number or name each stairway for easier identification. You will then move on to a description of all elevators located within the structure. Include construction, location, size, capacity, floors serviced, power switches, emergency power source and location, phone, key location, operating mechanisms, critical data from data plate, emergency accesses and escape hatches, and doors.

d. Go into more depth with any **physical barriers** other than doors located in the mission site/facility that could be used in a defensive posture during an incident. Include location, construction, dimensions, operating mechanisms and SOP for its use.

e. **Areas of limited access** located within the mission site/structure need to be identified. These areas will normally include the VIP quarters or offices, security operations center, facility operations center, business or corporate offices, finance offices, etc. Include their location within the structure, construction, security system (both active and passive), local communications capabilities in each area and existing emergency exits.

f. **Holding areas and safe havens**: A detailed description of the designated holding area for the client if one is required and all possible and designated safe havens within the mission site and related venues to include communications capability from each site. Rest room facilities, food, water, protective equipment and medical supplies in the safe haven should also be listed.

g. **Any weak points/areas inside the mission site/ facilities**: You must include a detailed description of all the weak points in the facility itself, including the location, nature of the weak point, and what advantages or disadvantages this area provides to the security detail, and alternate courses of action to bypass or negate these areas. As with anything you write for others, you need to provide as much visual information as possible so when submitting a report, the use of sketches, photographs, and video footage

is necessary to allow people who have never been to this facility before to have a greater understanding of it before the mission and during the planning phase.

5. THE ROOF:

A good **general description of the roof**(s) located on the mission site should include information about the type of construction, secondary structures (elevator rooms), location and type of antennas (high voltage), and any other obstacles or hazards. Then you need to cover all entrances and exits onto the roof including any fire escapes and ladders with the type of construction, location, number of steps/rungs with shape and dimensions, access, and destination.

6. COMMON SYSTEMS AT THE MISSION SITE:

a. Host site security:

(1) **Other security detail/bodyguards** if known. You need to find out if there are other VIPs or personnel who have their own security details, including the number of personnel on the detail, shifts, arms, equipment, and communications (normal and emergency). Building good rapport with other details working the mission site at the same time is vital.

(2) The **onsite security or security that's been contracted** to help with the event at the mission site. You should give an outline of the organization, equipment, arms, shifts and communications (normal and emergency), and number and locations of posts. Building rapport with them will help you in setting up your security for your client. If they are experienced and competent it may free up some of your resources for other areas in need.

(3) **Law enforcement.** Give a description of how any local or national law enforcement is involved in the day to day security of the site and how they will be involved during the mission if they are to have any regular security role. You should include shifts, arms, radios, and any other equipment, plus the POC and contact information. Ensure you get the correct rank and title for the POCs.

(4) The type and amount of **permanent security equipment on site** is important, so an inventory of all permanent security equipment should be conducted with the permission of the host site security manager or operations manager. The inventory should include any type of arms, amount and ammunition, CS dispensers, defensive equipment, locations of arms rooms, security measures, and any other information. If they have secure permanent arms room and weapons storage facilities you might be able to negotiate keeping a cache at the site for future operations and planning.

(5) With permission of the host site manager review current **plans for emergency response/defense and evacuation** in case of emergencies, both natural and manmade. This information should be available from the host site security and/or operations manager, and it should include all venues and agendas for the mission site. If there is none available then come up with an emergency response plan and evacuation plan for the client, and offer to develop one for the host site. Developing their plan can help you with yours and you will be able to avoid any conflict with the needs of the client. This will also build great rapport with the host site.

(6) **Mission site technical security measures**. This will include all cameras, sensors and alarms used outside the structure for daily security operations. You should include brand name, location, type, power switches, key locations, and bypass information. Make sure you know which systems are passive and which are active. Know and sketch the areas of coverage and dead spaces for all technical security measures outside the mission site, and where and how they are

monitored and by whom.

(7) **Mission site communications capabilities**. Discuss the communication systems used at the mission site by the host and all other entities that use the facility. This should include both mobile and stationary systems. Be sure to find out about normal and emergency local communication equipment, frequencies, location of the communication room if there is one, location of communications nodes, computer capability, telephones, faxes, power sources, POCs that maintain each item of equipment or the POC for communications at the site. You also need to test the ability of both the security detail's primary and back up communications to include cell phones at the mission site and any other areas that will be visited. If you do not have the same cell phone as your client you will need to test his for the ability to operate at this site.

(8) **Vehicles and other transportation organic to the mission site**. Review and provide an inventory of all vehicles used by the host at the mission site, including any tenants at the site and vendors. You also need to know normal parking or staging areas for these vehicles. This is important so the security detail will be able to pick up on vehicles that are out of place even if they have the correct markings.

(9) Detail the **fire protection systems** in place such as alarm systems, sprinklers, extinguishers, fire hose stations inside the facility and hydrants outside the facility and provide the location of nearest fire company, including what type of equipment they have, their capabilities and their reaction time.

(10) **Power supply** for the facility. What is the power source for the facility and what type of power system used in both normal and emergency situations? Where are the sources located, what is the location that power enters the facility, is it underground or above ground, location of the closest transformer or outside node that can shut down power to the facility, and locations of generators, power switches, fuel tank locations for on

site generators plus their fuel capacities? How long will the emergency power last on site? What areas are powered by the emergency power source and what are the emergency system start up procedures, or is there an automatic cut in?

(11) **Interior lighting**. What type of lighting? If there are different types, what are those locations? When is normal lighting turned on and off, or is it on all the time? The degree of illumination, locations of emergency lighting, what activates the emergency lighting, how often is it tested, how long does it last, are there any automatic lights, time delay lights, and what are those locations?

(12) Type of **air conditioning and air circulation systems** present in the building. What are the power sources and where are the on and off switches? Are the air circulation ducts located outside (can someone enter gas into the system)? Are the ducts large enough for people to hide in anywhere in the facility? Does the system work when there is loss of power and what is the average temperature in the building when the systems don't work?

7. PERSONNEL STRUCTURE:

a. **Key Personnel Data sheets**. All major players at the mission site, operations manager, security manager, protocol coordinator, maintenance supervisor, day and night supervisors should be listed. You need to list all personnel who have knowledge of the building, anyone who has access to the key control area or has access to the building. Include their work, home and cell phone numbers and home addresses, with strip maps and estimated response time to the mission site in case of an emergency or incident.

REMAIN OVERNIGHT (RON) HOTEL SURVEY

1. I**dentify hotel management** and coordinate for the visit with the hotel manager or their designated POC.

2. **Identify hotel security** since most hotels have their own security personnel. Find the security manager. It is always advisable to build a good working relationship with the proprietary security personnel because they can be valuable assets. They know the ins and outs for that hotel and the surrounding area and can recognize what is "normal" vs. something out of place. They can assist you in many ways such as finding secure parking for your convoy, traffic control to allow better in and out movement of the vehicles and locking down the elevators for moving the client and security personnel. Since they are familiar with who is at the hotel they can spot the difference between guests of the hotel and people who are not guests but loitering around. If they are former law enforcement or law enforcement working a second job they can be valuable because they can provide additional security, get law enforcement support on short term notice, introduce you to law enforcement POCs, conduct criminal records checks if needed and get assistance with traffic control when necessary.

3. The advance will be responsible for making **room assignments for the clients, security operations room, the security detail and any other considerations**. The advance team should know what the clients will need and what they like, but you need to keep in mind the security aspects of the assignments of rooms. The security operations room should be close to the client's room. The advance should obtain a floor plan or sketch one out and write the names and room assignments for each person. These sketches should be distributed to the detail leader and shift leader. The security operations room should have the mini bar sealed from use. I usually try to get a suite for this so the shift leader can sleep in the same location as the security ops center in case of an incident. If possible you should coordinate to get an Explosive Ordnance Detection (EOD) dog support and TSCM sweeps on all the rooms if the reservations had been made for a long time and it was known foreigners were staying there. It also depends on the operating environment and resources available. If it is a random selection of an overnight place, or it was not known who the reservations were for this may not be necessary, but once the rooms are obtained and cleared security needs be placed on them whenever no one is in them for the duration of the visit.

4. All hotels have some type of **services** that they offer: room service, maid service, massage, clothes washing, dry cleaning or porter service. Any room service ordered by anyone with the party should be delivered to the security operations room. The person on duty will sign for it and let whoever ordered it know its ready. As a courtesy, if the client orders something the person on duty can have a detail member deliver it to the client's room. Depending on the operating environment, once the rooms for the detail, including the advance and the clients, are obtained and cleared only security personnel are allowed in the rooms. Maid service should be discontinued if it is going to be a short stay or they should be escorted into the room when it needs to be cleaned.

Find out the type of services the designated hotel has or does not have and annotate it on the RON Survey, so everyone will know what to expect on arrival. The advance needs to identify the hotel's utility capabilities and whether the hotel has equipment for emergencies, such as a generator. If there is a back up generator, how it comes on (automatic or manual), how long before it engages, and the last time it was tested should be noted. It is also important to determine if the food and water at the RON site is potable and edible. If the kitchen is substandard and the water is not potable then coordination will have to be made to have these items brought in.

5. The advance also needs to establish contact with and develop rapport with the **staff of the hotel** and any other businesses located at the RON building. This includes not only the management and security as talked about previously but hall porters, stewards, doormen, restaurant managers and staff, floor waiters, maids, desk staff, etc. These are the people who are usually more aware of what is going on at the hotel at all times, and will be the first to notice someone or something that is out of place or unusual. These people should be told to contact the PSD team members if they see anything out of place or unusual before the arrival of the clients, during the stay and after the departure of the client. Of course how you handle this will be based upon the type of security profile you are using.

SITE SECURITY

The security detail including the advance should have security at the RON site or hotel broken down into 4 major areas of concern:

1. The first will be security of the **area immediately surrounding the hotel or RON site**. All buildings, roofs and natural features that overlook the hotel entrances, public areas, and parking areas must be checked to the best of your abilities. If possible vet the occupants or, if that is not possible, find out how long they have been occupying those spaces, their background, etc. from local sources. If they have been long term occupants the risk is small but if they are new occupants who have moved into those spaces after the travel plans where made for the client then you should put a spotter to watch those spaces for any activity that is suspicious before, during and after the client's departure. If possible, photograph the occupants so you can compare them to other photos or footage of missions when looking for threat surveillance. You should also consider using counter surveillance teams along the routes to the hotel the client will be staying in to look for possible threat surveillance or attack preparations and possibly staging a quick reaction force in another location along the route and close to the hotel.

2. Second will be the **security of the hotel or RON site** itself. When the situation requires it, the physical building the clients will be staying might need to become a 100 percent defensive area, which means all entrances into the building will be under the security detail's direct observation and if possible control. This may not be possible depending on your resources so at a minimum need to control the floor the client is on, and any entrances or exits they will use during their stay, especially during their time of arrival or departure at the location. This means controlling the area around these entrances and exits also. You will need to have people at these entrance/exit points, inside and outside, in plainclothes doing area observation to look for threat surveillance of any type and anyone who looks out of place or suspicious. This should be done at least an hour before the client's expected departure or arrival time.

The departure and arrival points are always an area of great risk for you and the client because he is outside any armored or protected areas, on foot, while the vehicles are stopped waiting for him. Everything is moving slowly and these points are usually known to everyone, so it's not unusual to have people already there waiting to see an important person, to beg for money, etc. The threat likes areas like this because the client/target is between cover. He is not surrounded by walls and/or a vehicle but is in the open. This is the best time for the threat to hit a target because it is the safest and least resource intensive to them and whatever they do will cause a disruption.

The client should not move from his room or arrive at the arrival/departure point until the all clear is given. You do not want him hanging out in the lobby waiting for the area to be checked and cleared or for the vehicles to arrive. You also don't want to pull up to the arrival point if there are vehicles or pedestrians blocking the road and/or the entrance. Keep the clients in that safe mode until they can get there and get in or out as fast as possible with no delays.

When reserving rooms make sure to get rooms on each side of the client room, directly across from the client room and the one above and below the client room. If there is more than one client do the best you can. Having rooms in this arrangement will allow you to have some measure of control around the client if necessary.

The threat of FIRE is a very real threat in most locations that is very often overlooked or not considered. The advance should make sure that the survey they conduct includes a detailed section on fire probabilities, and fire fighting capabilities of the site to include escape routes from all areas that the client may use during his stay, the sites main evacuation routes, designated assembly areas, serviceability of emergency alarms and emergency lighting and fire fighting equipment located on the site. When the detail arrives at the location it needs to be briefed by the advance on what actions will be taken in case there is a fire, then the client or clients will be briefed separately on the actions they need to take.

3. Thirdly will be the **security of the room or rooms, tents, or the areas where the clients and security detail will be sleeping**. The advance needs to make sure that the areas obtained for the clients and the security detail will mitigate the possible threat for that area. Rooms should face inward with no windows facing the street or on the outside of the building. They should not be above the 7th floor of the structure since most fire fighting equipment does not extend that far. They should not be on the ground floor since they will be too accessible, and you have the threat of secondary blast effects on that level. Also you want rooms high enough so objects cannot be thrown at the balcony or windows. The rooms should all be located at the end of a hallway away from vending machines, ice machines and elevators, to lower the amount of foot traffic by the client's room and the security detail's rooms, since there will be no reason for anyone to be in that area. And you need to avoid rooms on floors that have common access areas such as bars, gyms and swimming pools.

4. Lastly, you need to have **security in place within the structure or venue when the client is moving** to different areas so basically security for anything taking place within the RON site. This will be necessary even if the hotel is totally secure. The clients need to have their bodyguards or detail members with them whenever they are moving around the perimeter of this type of secure area. All meals, if not taken in the room, should be taken in the largest restaurant in the venue with a table selected by the detail leader. The table needs to be away from windows, close to, but not right next to, any doorways or other exits, and situated so other diners do not have to pass it to exit the facility or use the rest rooms. The clients should sit with their backs against the walls when possible. Never use the same table even if you have to use the same restaurant because of a long term stay.

A thing that tends to happen to security details when they are going from hostile areas/high threat areas or war zones to a semi-secure location that has some aspects of civilization such as nice restaurants and bars is they drink. No detail member who is carrying or has access to weapons should ever drink alchohol at any time, even off duty. I don't trust people who are drunk with guns around me when I go hunting, why would I want to in a threat environment? In the public areas, if the client needs to use a rest room, a person goes in with him and one waits outside the door at a minimum. You don't have to follow him into a stall but someone needs to be in there in case of an incident.

GENERAL NOTES

DINING OUT: When eating out at public places the detail needs to order a meal that will not take a lot of time. If they know the place, they should call and pre-order and ensure to order takeout for those in the vehicles. When eating in, make sure you sit where you can observe the clients and any entrances and exits. Talk with the manager or head waiter to make sure he knows why you are there, that you need to have a rush order, and you need too pay as soon as possible. Do not worry about dessert because the detail needs to be finished and ready to go when the clients are. You do not want the clients waiting as you finish your meal or are paying for it when they are ready to leave. In some cases, the detail might not eat at all. I have had details bring sack lunches and the members inside just had tea or coffee while waiting for the clients. This way they could observe more instead of doing two things at once. Make sure you have a form of payment the site will accept because some places do not take credit cards, only cash, and some will only take local currency. Finally, one detail member pays for everyone. If you do not have a fund for it, you can divide the check later after the mission.

LUGGAGE: The client's luggage and the detail's luggage need to be guarded and under the detail's supervision the whole time. Do not let bellhops move luggage to the rooms because you won't know if they do something to it or not. If you do let them move it then the person in charge of the baggage detail will be with the luggage and equipment the whole time until it is in the rooms or stored back in the vehicles. Know the luggage count so nothing can be added by any local support staff, either going to the rooms or into the vehicles.

VEHICLES: The vehicles used by the detail need to be closely watched and guarded. This is one of the most vulnerable areas of security that you and the client will have. The vehicles need to be placed in a secure garage and locked at all times when not being used. And by secure garage I mean one that has restricted access and guards. If this is not available you might consider running a guard rotation yourself or putting the vehicles under video surveillance. Whenever a vehicle has not been under the direct observation of the detail you will need to conduct a vehicle bomb search on it prior to movement. You are also going to be looking for tracking devices and/or technical surveillance devices such as microphones that might have been placed on a vehicle.

THREAT/SECURITY LEVEL: The level of security you need to use as with everything else you will do depends on the threat level you have assessed for that location, the type of profile you are using, length of stay at the facility and the activities that are taking place. Doing temporary security at away locations can be a very difficult task so good planning, coordination and use of your available resources is vital to protect your client and yourself.

AIRPORT SURVEY

When using military or civilian airports that have either a main terminal or private aviation hangars, the advance first needs to get a copy of the travel itinerary. If the client is arriving at a main terminal on a commercial airplane, you need to coordinate with airport security to meet the client at the gate, if possible, however some airports only have one gate, and the arriving client might have to go through customs. If so, you need to wait for him to clear these areas before meeting him. If he is traveling with a bodyguard from your company there will be no need to identify yourself so just wait for them, greet them then move them to the vehicles. If the client is traveling alone and you do not know him, make sure you have a picture of him, so you can identify him and go introduce yourself. Do not have a sign with his name as this will let everyone there know who he is and he could become a target just for having someone meet him.

After you introduce yourself and get his bags, move him to the vehicles for transport. Some VIPs and/or clients that fly into major airports have company or private planes that will land and arrive at a private or executive aviation type hanger which is away from the public hangers. If this is going to happen make sure you know which hangar and be there to meet the client.

You need to obtain the aircraft tail number, mission number, flight number and type of aircraft that the client will be flying on or as much of this information as possible from the either the client, the security detail or the person making his or her travel arrangements prior to the client departing his location. By having this information beforehand you can verify that the airport has the same plane information and that it is scheduled correctly. If not you will have to find out who is wrong or what has changed and may have to make coordinate with the airport or the client's office or the security detail about the changes. This type of information also helps in identifying the plane when it lands.

When conducting an advance at an airport, you should try to obtain access to the tarmac to greet the client and move him straight to the vehicles, however this may not be possible when arriving into a new country or if he has lots of baggage and needs to pass through customs. Some airports may require a special pass or escort to drive on the tarmac to pick up the client plane-side. Once again, this is where hiring an expeditor, who has the necessary knowledge and contacts to help you get access to these restricted areas, is a good idea. If you cannot get access, the expeditor may be able to. NOTE: In Jordan, several companies hired expeditors to meet arriving people and walk them through the immigration, baggage claim and customs process, then straight to the vehicles. Many times depending on the length of the stay the expeditor had already obtained passes for his clients to bypass lines.

If you are meeting an aircraft on the tarmac you may also need special equipment like hearing and eye protection.

The advance should find out all requirements for passing through customs and immigration areas for the airports the client will be flying into. In some countries you can coordinate with customs and immigration officials, especially if your client is doing work for the host nation. Have the necessary papers ready and any other necessities that may be required to get your client through the airport smoothly. NOTE: Some countries will only allow you to purchase an entry stamp with that host nation's currency, so having this for the client to get through the process quickly will be a big boon to your security. Some places you may have to give gratuities or bribes to help smooth the process do what is necessary to maintain security.

It is always a good thing to find and coordinate the use of any VIP or first class lounges for the client because sometimes they like to stop in the lounge prior to departing the airport if they are going straight to a meeting from the airport or they might want to use the lounge while waiting to leave the country.

If the client is coming by private or company aircraft, you will need to obtain secure parking for the aircraft so if it has not already been coordinated, contact the airport to obtain it. Some airports, depending on their location and resources, may not have a secure area. It is the advance team's responsibility to ensure that the aircraft is secure during the time of the visit, which may require hiring a local static guard force depending on the resources and manpower available to you.

The advance needs to find out how the baggage handling system works for the airport. All client and detail bags should be marked with identification and a marking (I used to paint locks purple with one number and place it on each bags handle) that will be easily recognizable to the person responsible for the baggage detail. This way he will be able to recognize all bags and collect them without the client standing around in an unsecured environment. The person responsible for the baggage detail also needs to know the quantity of the bags coming in to ensure he has the correct amount. The security detail assumes responsibility for the security of all bags for the entire trip once they are in the detail's possession. That includes moving them to and from the hotel or any other locations visited during the mission. You should never leave the bags unattended and always remember to place the client's bags in the vehicle last so they can be removed first when you arrive at mission site or overnight site.

When a client is taken to the airport for his departure the advance should know what will be necessary to get him through the process as quickly as possible. Once again hiring an expeditor might be necessary. Once he is through the check in process and to the gate area where security cannot usually go, knowing when the plane will go wheels up is important. It is necessary for the security detail to know this so they can stay at the airport until the client is actually off the ground and on his way. I would have the client call me when he boarded the plane then watch it until it took off. In some cases the flight might be delayed or cancelled and you would not want the client to be stuck at the airport without security available for him. In some places flights are pushed back many times and the detail leaving the airport area may take awhile,

especially in a high threat area or a place with bad weather or limited runways or tower capabilities.

Other things that might be necessary depending on your location:

- Highway security for aircraft overpass on final approach

- EOD runway and terminal sweep

- Accessibility to the tarmac for the convoy

- Types and locations of barricades on tarmac or in terminal (crowds, public access)

- Convoy staging area

- Normal routes on/off tarmac, in/out of terminal, and emergency evacuation routes

- Any special coordination with the airport manager or the security manager to ensure smooth a smooth operation of retrieving the client from the airport and ensuring his safe departure

- Have the emergency and fire crews at the airport been trained and do they have the necessary equipment to respond to an incident?

HOSPITAL SURVEY

1. Is it a civilian or military or other type of hospital? What services does it provide - inpatient with emergency room, inpatient without emergency room, outpatient with emergency room or outpatient without emergency room (ER)?

2. Identify the primary POCs for the facility, including the on-duty charge nurse located in the emergency room, security manager or supervisor, operations manager, etc.

3. What are the medial capabilities at that facility? Does it have x-ray capabilities, whole blood or specifically the client's blood type on stock? Is there a cardiologist and a trauma unit, a pharmacy? What are the hours of operation for each? If they do not have what you deem is a necessary function then find out what hospital in the area does, in case you have to move the client to that hospital because of a specific medical emergency.

4. Find out if the hospital has a VIP room or private rooms available and their locations.

5. Determine the onsite security force capabilities for the hospital. Find the facility security manager and coordinate with them. Determine the capabilities and availability of the security force and if they have procedures in place for security operations when a VIP, foreigner or high profile patient stays there.

BASIC PHYSICAL SECURITY CHECKLIST

1. IDENTIFICATION:

    a.   Local and official name.

    b.   Functional description.

    c.   Local address.

    d.   Map reference.

    e.   Geographic coordinates.

    f.   Additional information.

2. BUILDING DESCRIPTION:

    a.   Identification.

    b.   Type of construction material.

    c.   Type of roof.

    d.   Blue prints.

    e.   Floor plans.

    f.   Description of roof.

      (1) Entrances to building.

      (2) Skylights.

      (3) Air conditioning ducts.

      (4) Maintenance accesses.

      (5) Elevator shafts.

      (6) Emergency exits and fire escapes.

      (7) Ventilation Systems.

      (8) Ladders.

      (9) Additional information.

g. Entrances to building.

  (1) Main entrance.

  (2) Other entrances or exits.

  (3) Non-standard access points.

h. Sewage and drainage Systems.

i. Water system.

j. Interior description of building.

  (1) Floors.

  (2) Corridors.

  (3) Doors.

  • Locking mechanisms.

  • Hinges and placement.

  • Inside dead bolts.

  (4) Windows.

  • Locking mechanisms.

  • Type of glass.

  (5) Locks, keys, who controls
      and has access to them.

  (6) Stairways.

  (7) Elevators, serviced by, key controlled by.

  (8) Physical barriers.

  (9) Lighting, back up power for lighting.

k. Active security.

  (1) Company security guards.

  • Posts.

  • Arms.

  • Ammunition.

  • Riot control devices.

  • Additional information.

  (2) Contract watchmen.

  (3) National police agents.

  (4) Alarm systems.

  • Closed circuit television.

  • Sensors.

  • Radio backup.

  • Monitored where and by who.

  • Back-up power systems.

  (5) Additional information.

I. Communications equipment available.

  (1) Model.

  (2) Type.

  (3) Number.

  (4) Location.

  (5) FREQ/Channels.

  (6) Antennas.

  (7) Remarks.

  • Telephone system.

  • Emergency lighting and power system.

  • Additional information.

3. DESCRIPTION OF GROUNDS:

    a.   Structures.

    b.   Entrances to grounds.

    c.   Perimeter fence.

        (1) Gates, type, secured by.

        (2) Posts/gates secured in ground by, depth

    d.   Terrain

    e.   Vegetation.

    f.   Lighting-external.

    g.   Map or sketch.

    h.   Power entering grounds.

    i.   Water supply entering grounds.

    j.   Sewer lines.

    k.   Additional information.

4. DESCRIPTION OF SURROUNDING AREA:

    a.   Possible observation points.

    b.   Relative distances of key terrain.

    c.   Additional information.

5. KEY TERRAIN:

    a.   Predominant terrain.

    b.   Drops zones.

    c.   Landing/Pickup zones.

    d.   Airfields.

    e.   Critical lines of communications.

    f. Additional information.

6. MEDICAL CONSIDERATIONS:

    a.   Embassy medical staff and facilities.

    b.   Civilian hospitals.

    c.   Additional information.

OPTIONS FOR VEHICLE SECURITY

1. Register vehicles of high-risk clients and the security detail with the **local vehicle registration**.

2. Use **host nation license plates** for all vehicles.

3. **Modify exhaust pipes** by placing mesh over them to prevent insertion of explosives.

4. Initiate a **vehicle/personnel tracking system** for each client.

5. Install **vehicle tracking systems** in all vehicles used.

6. If vehicles are unarmored, acquire **ballistic plates** that can be easily installed and removed from vehicles.

7. Require all clients and security personnel to have a **cell phone** or other means of communication.

8. Conduct **route surveillance** in advance of convoy movements.

9. **Rotate vehicles** as often as possible depending on resources available.

10. Have **removable vehicle decals** for areas that require them to gain access.

11. Develop a listing of **local safe havens** and provide it to each driver.

12. Train all personnel in **evasive driving tactic**s and techniques.

13. Train all personnel in **detection of surveillance** and actions to take.

14. Train all personnel in medium and high threat areas on **vehicle searches**. Use realistic training aides.

15. Issue **flashlights and inspection mirrors** to all vehicle operators.

16. Establish a **duress code system**.

17. Establish **checkpoints** along all routes used in the area of operations so personnel will report their security status at these checkpoints.

18. Install **locking gas caps** and gas doors on all vehicles.

19. Install **alarm systems** on all vehicles.

20. Develop a **pre-operating checklist** for all vehicles. This list needs to include bomb search for each specific vehicle type.

OPTIONS FOR PERIMETER AND ACCESS

1. Have garbage trucks enter the site empty to make it easier for inspections.

2. Place barriers at unused gates.

3. Run aircraft cable through gates and anchor it to a concrete pillar or jersey barrier.

4. Use landscaping to mitigate high-speed approach routes.

5. Use sensors and/or CCTV at secured gates.

6. Use high-pressure hot water to control foliage on the perimeter.

7. Assign escorts for delivery vehicles.

8. Install card readers at unmanned gates that require infrequent entry.

9. Have local law enforcement patrol the exterior perimeter.

10. Use motion activated lights in areas of approach and un-patrolled areas.

11. Use sensors in likely avenues of approach on the perimeter.

12. Randomly man observation points or posts to overlook the perimeter.

13. Verify freight deliveries through bill of lading or telephone verification.

14. Verify deliveries of food and household other items through a list provided by the site or facility.

15. Man vehicle checkpoints throughout the site during open hours.

16. Cable jersey barriers together to increase their effectiveness

17. Outline where barriers are to be placed. Take a photo of the barriers in place.

18. Limit the delivery hours to the site and installation.

19. Establish an over watch position for the entry control points.

20. Provide portable duress alarms for entry controllers.

21. Install CCTV at entry control points.

22. Board all buses to conduct ID checks when entering the site.

APPENDIX
PERSONNEL SEARCH TECHNIQUES

GENERAL SEARCH TECHINQUE

Position the person being searched out from a wall (or car) with their legs apart and their hands against the wall in a leaning position, in such a way that he cannot move without falling down, or can be easily knocked over.

a. The searcher should always **work from behind**.

b. **Two searchers should be employed**, one searching and the other covering.

c. All searches are conducted in a **business-like manner with conversation limited** to requests/instruction necessary for conduct of the search. Extend proper respect to all personnel being searched; the aim is to provide security without creating animosities which could develop into trouble in the future. Do not exceed your authority.

TYPES OF SEARCH

There are two types of search: a quick body search or frisk or a detailed body search.

QUICK BODY SEARCH OR FRISK

The frisk is used either as a preliminary search to detect weapons, or as the usual from of search in a low threat environment.

1. Follow a logical sequence from head to toe. You should have a special room or area set aside for this for several reasons. First for the extra security because an incident in a separate space is easier to control. Second it separates a possible threat from the people around him or her. When searching, you should use both hands and stroke (rather than pat) all clothing. If you have the resources, for quick body searches use a metal detection system.

2. The following areas should be carefully checked:

- Hair and in or under hats.

- Armpits.

- Inside legs.

- Groin or crotch area.

- Half-clenched hands

- Any medical dressings.

- Any bags or cases carried.

- Walking sticks, umbrellas, crutches, etc.

- Shoes/boots.

Conducting a Detailed Body Search

You should have a special room or area set aside for this for several reasons. First for the extra security because an incident in a separate space is easier to control. Second, it separates a possible threat from the people around him or her. Third, it allows some privacy for the person being searched. Females should always search females; males should always search males, even children. In some countries and cultures this is vital and is a major insult to the people if not followed. You should follow the sequence below for the detailed body search:

a. Establish identity. (What gives you the authority to do the search?)

b. Establish ownership to baggage. (Has to be positively identified.)

c. Invite the person to turn out and empty all pockets in clothing he or she has with them.

d. Invite the person to remove all clothes, jewelry, watches, etc.

e. Inspect body from head to foot, paying special attention to hair, ears, mouth, teeth, body orifices, crotch, groin, between toes, etc.

f. Examine clothing, paying particular attention to linings, seams, buttons, belts, shoe/boot soles and heels, etc.

g. Examine contents of pockets.

h. Examine baggage and other articles (sticks, umbrellas, etc.)

ALWAYS REMEMBER

a. *Women must search women; men must search men.*

b. Watch for facial reactions, nervousness, or sweating.

c. Work in pairs and search each individual separately.

d. Be courteous.

e. You must know the customs and cultures for the area you are operating in. Making a mistake, degrading someone or embarrassing someone could have severe consequences during any operations you will be conducting in the mission area.

APPENDIX
ROUTE SURVEY FORMAT

1. The title page should include the following details:

   a)   Survey team members.

   b)   Level of assessed threat on the route.

   c)   Date and time the survey was conducted.

   d)   Date and time of planned mission.

   e)   References:

   i) Maps:

   • Map sheet numbers.

   • Map titles.

   • Map folds.

   • Scale.

   ii) Aerial photographs, satellite imagery.

   iii) Digital photos and video.

   iv) Street maps.

   v) GPS used and datum.

   f)   Vehicles used, high or low profile, should be same type used on mission.

   g)   Traffic conditions.

   h)   Any additional relevant information.

ROUTE DETAILS

2. Any details regarding the route such as:

   a) Embus/Debus Point. Description, diagram and/or pictures, security measures, vulnerable areas.

   b) Main Route. Including route sketch, safe havens, refueling, feeding shops, toilets, hospitals, telephones, emergency services, choke points, danger areas, route maps.

   c) Alternative Route. As above.

ASSETS REQUIRED

3. Internal – resources required to carry out the task.

4. External - all assisting forces (police, military, medical, fire, recovery).

5. Hospitals - grids, addresses, telephone numbers of accident and emergency POCs.

6. Safe havens.

7. Landing sites for air support if available.

SERVICE SUPPORT

8. Uniform of detail, high or low profile, etc.

9. Equipment - including medical equipment.

10. Rations required for mission.

11. Weapons required ammunition.

12. Transporting requirements, vehicles, support, etc.

COMMAND AND CONTROL

13. Communications:

   a) Dead spots along route for
      radios and cell phones.

   b) Communications with supporting
      agencies and your Ops Room.

   c) Backup systems to include satellite phones

ANY ADDITIONAL INFORMATION

14. Any unusual occurrences during the survey that
could affect the mission.

APPENDIX
SECURITY DETAIL OPERATIONS ORDERS

The aim of briefing operations orders to the security detail for each operations is to ensure that they and any other details working with them know what is going on, including all the aspects of the planned operation that will pertain to them and their section, what their task will be for this operation and how they fit into the overall security plan for the operation. It is important that you write out an operations order, and warning order for the mission because no matter how experienced you are there is the chance you might forget something or someone or leave out a vital detail.

There are two types of orders I will discuss: warning orders and operations orders. A **warning order** is issued to everyone that will be working on the mission. This permits them to start the preliminary planning necessary for their portion of the mission. A warning order should include the following points:

- General outline of the proposed operation.

- Delegation of tasks.

- Time Limitation.

- Time/Location for giving the operation orders.

Operation orders will be the detailed written orders that are given once the operational plan has been formulated.

PROTECTION ORDERS EXAMPLE FORMAT

Sequence:

- Situation

- Mission

- Execution

- Service support.

- Command and Signals.

VISUAL AIDS

Visual aids should be used during the briefing including maps, sketches, aerial photos, satellite imagery, and/or videos.

FORMAT

PRELIMINARIES

• **Task**. The summary of the operation will include:

1. Area.

2. Reason for operation (purpose of trip).

3. Political background of local area.

4. Brief sequence of events from mission start to finish.

5. Duration of operation.

• **Ground**. The ground is covered in two phases:

1. **Phase 1**: A general outline of operational area.

   • Type of area - rural / urban and density.

   • Road system / rail system / subways.

   • Emergency facilities to include hospitals / police stations / military bases.

2. **Phase 2**: Detailed description of specific locations of interest:

   • Venue, facility and or static site to be visited.

   • Embus / debus points.

   • Routes:

     o Main / alternate.

     o Choke points, danger areas.

     o Safe havens.

   • Weather: climate (if applicable) weather forecasts, first / last light, moon state (if applicable).

SITUATION

• **Threat**

1. Level and type of threat.

2. Known enemy personnel.

3. Past operations and tactics.

4. Weapons.

• **Local Population**

1. Politics/Cultural background.

2. Language

3. Attitude of local population towards the clients and/or client companies.

• **Friendly Forces.** Police and military units within the operational area.

1. Locations of units.

2. Restrictions on the detail imposed by military and police.

• **Attachments.** These can include:

1. Advance.

2. EOD/tech surveillance search teams for the site.

3. Fixed site counter surveillance.

4. Counter assault team or security advance patrol.

5. Site liaison or POCs.

6. Local law enforcement escorts.

7. Air support.

MISSION

Clearly define aim: to allow the client to do their job and conduct their mission without incident but "to protect the client" at all times.

EXECUTION

General outline:

- Phases of the operation
- Outline activity of each phase

Detailed tasks: Detailed tasks and essential information is given phase by phase.

Phase 1:

- Grouping (SAP) - task
- Grouping (BG) - task
- Grouping (Detail) – task

Rules of engagement: These will include when it is okay to fire your weapon, and what the escalation of use of force will be. This must be concise and fully understood by all detail members taking part in the operation.

Action on:

- Main route - not useable regardless of reason.
- Breakdown of main body vehicle.
- Accident regardless of vehicle.
- Illness, type, how serious.

- Verbal / physical assault or kidnapping attempt.

Safe haven.

- Location of each safe haven, description.
- Actions to be taken on arrival at a safe haven after an incident.

Coordinating instructions: These can include but are not limited to:

- Rehearsals.
- Preparation of equipment / vehicles.
- Communications checks on all systems even back up systems.
- Meals laid on at site, carried in vehicles.
- Summary of events and timelines.

SERVICE SUPPORT

1. **Uniform** for detail to include and special clothing requirements for the bodyguard or detail leader at the facility or event.

2. **Arms / ammo.** This includes ammo for personal and support weapons.

3. **Pyrotechnics** This includes smoke and flash bangs.

4. **Vehicles** This includes client and escort vehicles. Things to consider:

- Baggage, how much and in which vehicle or a separate baggage vehicle.

- Refueling point.

5. **Keys.** Key control and locations:

- Vehicle.
- Safe house.
- Venues/Facilities if applicable.

6. **Maps.** Maps / GPSs needed for :

- Planning.
- Navigation.

7. **Documents.**

- Special passes.
- Security gate cards.
- Weapons permits.

8. **Personal equipment.**

- Change of clothing.
- Overnight stay.

9. **Team equipment.**

- Search equipment.
- Viewing aids, binoculars, camcorders, cameras.

10. **Accommodations, safe rooms in facility.**

11. **Medical.**

- Vehicle medical packs, and personnel med kits.
- Hospitals.
  a. Location.
  b. Routes.
  c. Telephone numbers and POCs.

12. **Air support.**

- Type.
- Location.
- Type of standby.
- How to contact.
- Reaction time.

COMMAND AND SIGNALS

a) Chain of command.

b) Radio types, compatibilities.

c) Low visibility attachments.

d) Batteries, chargers.

e) Frequencies.

f) Call signs.

g) Codes / nicknames.

h) Contact schedules.

i) Lost communications procedures.

j) Air communications procedures if applicable.

k)   Contact telephone numbers.

l)   Synchronize watches.

THE DELIVERY OF ORDERS TO PERSONNEL

Thing to consider:

- Security of location.   The walls have ears.

- Distractions.   Avoid places where things will distract the orders group.

- People walking in.

- Cell phones going off.

- Radios too loud in room or in adjacent rooms.

- Vehicles passing by, honking horns, etc.

- Arrangement of room.   Briefing room should be:

  - Laid out prior to briefing.

  - Laid out in such a way that each detail will be in their respective groups.

- Delivery.   Orders should be delivered as follows:

  - Fluency - speak clearly and distinctly and in a language everyone understands

  - Speed - allow for note taking.

  - Format - follow a standard format.

  - Allow for questions.

CONCLUSION

The presentation of good, concise formal orders is paramount to the success of protective security operations. It ensures that every member of any security detail knows:

- The client's planned agenda in detail.

- The security plan formulated to cover the mission.

- The part each security detail member who is going on mission will play within the security detail.

APPENDIX
OBSERVATION

Here are some indicators of what you should be looking for when observing individuals and or vehicles during security operations regardless of the function.

SUICIDE BOMER INDICATORS

• Alone and acting nervous.

• Loose or bulky clothing which may not fit into current weather conditions.

• Exposed wires or cables which may be visible around the sleeve or the front of the garment.

• Rigid mid-section because having a rigid device or explosive vest makes it difficult to bend or he could have a rifle or other shoulder fired type weapon whose length makes it difficult to bend.

• Hands clenched in a fist or something being held tightly at all times. He could be holding the detonation device with a dead man switch which means once it is released it will set the device off.

• Not using two hands for normal activities, keeping one hand closed like he is holding something and using the other to drink or smoke.

• Men who have recently shaved all body hair, including eyebrows. This is a potential suicide bomber

POTENTIAL PROPS

Props are used to transport devices so they are not noticed, to make people seem harmless, or to help people blend into that type of environment. These can include but are not limited to:

• **Baby strollers**. Most people see a young woman or couple pushing a baby stroller and consider them harmless, but remember explosives and weapons can be stored in one. Do the other activities of the person pushing the stroller fit?

• **Shopping carts**. While they are normal in a lot of places, usually you don't see a shopping cart away from stores unless it is being used by the homeless or by kids to play with. What is in the cart? Who is pushing it? Should it be in that area?

• **Backpacks** are used by mostly by younger people or people who are former military or law enforcement or come from a similar background. While they can hold lots of materials, most people use them for transporting day to day things such as books for school. Seeing a person with a backpack, you need to try to discern how heavy it is. Are there wires running from it to the person carrying it? Is it over one shoulder or two? Does the person fit the profile for someone who normally carries a backpack?

• **Briefcases** are used by business people and others for work. They usually transport papers and other work related items which do not weigh a lot. When you see someone with a briefcase, does he look like he should have one? How heavy is it? Does it fit that person's demeanor? Does it fit into the area?

• **Musical instrument cases** are widely used in some areas of the world, but mostly seen around music facilities or schools with music programs. If you see someone or several people carrying these types of cases, do they look like they should have them? Do they look like musicians? Why are they in this area? Are they carrying them right, or in such a way to open them quickly?

• **Golf bags**. Golf is not really played everywhere or by everyone (I have friends that disagree with this), but weapons and explosives can be carried in these bags. When you see people with them, do they look like they play golf? Are the bags too heavy for them? Are the bags overcrowded? Is there a golf course in the area?

• Other types of **gym bags or sports equipment bags**. Why are they being carried? Gyms in area? Sporting goods stores?

LOCAL AREA INDICATORS

• **Suspicious vehicles or people hanging out in the same location**: People who may or may not have a reason to be there, people or vehicles that are always in a position where they can observe you or in a potential attack site should be considered possible threats. This includes beggars, vendors, food sellers, newspaper sellers, and roadside stands, especially if they are new to the area or started working or hanging out there AFTER you or the client established a presence in that area.

• A person or people sitting in **vehicles that allow observation** of you or the client: People who sit where they can OBSERVE you or the client either during movement or while at fixed sites are always suspicious. Also be suspicious of people who are using video or camera equipment from a parked vehicle that is not in a tourist location, people using high powered lenses, people who are focused on your movements or the movements or sites of the client, or people who are pointing at security, the client and/or a facility while looking at something in their hand.

A pattern or series of false alarms or prank calls: For the threat to gain planning information for emergency response times they sometimes call in false alarms or set off alarms to get emergency services to respond so they can time them. Against a facility such as a residence or work site, they can set off perimeter alarms or cause a disturbance to gauge how security reacts and their reaction times. They can also cause disturbances such as parking a car then setting of the car's alarm, or setting a fire in the street in front of a facility or near the gate or perimeter fencing to see how security responds to those incidents.

INDIVIDUAL BEHAVIORS INDICATORS

• The person does not fit the clothing style he is wearing.

• The type clothing is not appropriate to the weather.

• The clothing does not fit the demeanor of the area, i.e. work clothes vs office attire

• The clothing does not fit the person, i.e. the person is well groomed and clean but his clothes look shabby or dirty, or the person is poorly groomed and unclean but his clothes are well kept and clean.

• The clothing does not fit into the location, such as wearing heavy coats, rain coats or jackets in a building.

• The person does not act normally for the situation and clothes. i.e. walking into a building out of the rain but not unzipping his overcoat, as most people do this to shake water off.

The appearance of a person can give off indicators also. For example does he have white walls around his ears or pale skin patches where he has recently gotten a haircut or shaved off a beard or mustache to better fit into that environment?

VEHICULAR INDICATORS

Behaviors with vehicles can be indicative of surveillance and/or an attack:

• A driver who is focused on your vehicles and not paying attention to the traffic around him or you.

• Peeking, when a vehicle peeks around other vehicles to keep you in sight.

• Sitting at an intersection or a traffic circle, passing up chances to pull into traffic while watching your vehicle's approach

• Vehicles that have one or two male passengers when it is unusual in that area for vehicles to ride around half empty.

The key when observing for the unusual is to find out what is unusual for that area. Keep in mind the culture you are working in and the operational environment. If you cannot filter out the distractions and get to the things that give away a potential threat or can stop an attack then you are giving away one of your best chances at AVOIDING the threat.

ABOUT THE AUTHOR

Bob Deatherage spent over 22 Years in the US military. In 1979 he joined the US Marine Corps working for the Naval Security Group as a Signals intelligence, Electronic Warfare specialist. He left the service for several years, joining the US Army in 1986 working in the intelligence field. In 1987 he was assigned to the 1st Special Forces Group Military Intelligence Company and after serving 2 years as a Special Operations Team Chief, he was selected for attendance to the Special Forces qualification course where he earned his Special Forces tab. He then continued to work in 1st Special Forces Group located at Ft. Lewis, WA as a Special Forces Communications Sergeant, working in Southeast Asia. During this time he was selected to work in an other assignments in Washington, DC area. He left that assignment in 1996 and went to 7th Special Forces Group, located at Ft. Bragg, NC where he spent his first year training Mexican Police special narcotics unit, then had multiple missions to South America on various assignments.

He spent his last four years with the 2nd Battalion, 1st Special Warfare Training Group, his first yeas as an Operations NCO for ASOT, SERE and ATD while also working as a instructor for one year on the Antiterrorism Training Detachment and the last two years as the Non-commissioned Officer in Charge, before retiring and returning to his home in Saint Joseph, MO. Since retirement he as worked as a Security Manager for an international construction firm in Iraq, working first in Mozul, then in Baghdad, the biap and Camp Victory before leaving in 2006. Since then he has conducted training seminars for military special operations units and civilian law enforcement organizations throughout the United States on a wide variety of subjects in antiterrorism, security and PSD operations for KI International and Global One Resource Group.

Bob has a Masters in Business and Organizational Security Management and is a member of the American Society for Industrial Security.

Index

#

2IC 34
360 degree observation 30, 133
360 degree security bubble 29, 31, 79
4 wheel drive vehicle 155

A

accessibility 47
access requirements 176
advance 26–28, 31, 34–35, 49–59, 113, 116, 174, 175, 185, 189
advance leader 59
aerial surveillance 66
after action review 24, 37, 42, 54
agent in charge 33
airport 26, 58, 188–189
air circulation systems 183
alarm 168, 170, 193
alertness 75
ambush 121, 154
ankle holsters 29
anti-surveillance 67–68
areas of vulnerability 161, 180
armor 120
armored vehicles 123, 128
attack 45, 117, 152
attack recognition 31
audio surveillance 66
awareness 73–78

B

baggage 34, 188, 189
ballistic plates 193
barriers 179, 181, 194
behavioral indicators 77–78
blatant surveillance 62
blueprints 180
bodyguard 23–26, 29, 33, 50, 116, 152, 181
bomb threat 171
bullet proof windows 123, 127
bunkers 170–171

C

card readers 194
CCTV 194
checkpoints 193

chemical lights 171
choke point 70, 105, 179
client 21–22, 31, 35, 48, 90
client profile 39
close protection team 30, 34
combat life saver training 22
communication 24, 29, 58, 90, 106
communications scanning 66
communication systems 182
community surveillance 63
company profile 39
competing companies 61
concealment of weapons 29
conduct 23, 169
consecutive turns 68
construction 40
control points 178
convoy 65, 113, 116, 154, 174, 189
convoy equipment 131
coordinates 177
cordon 157
correlating movements 64
counter-attack 156
counter ambush drills 117
counter assault team 29, 34–36, 69, 153–157
counter sniper teams 155, 159–164
counter surveillance 69–72
covert profile 154
covert surveillance 62
CPR training 22
CQC 157
cranks 38
criminal threat 37
crowds 79, 81, 89
cul-de-sac 68
curbside 114
customs 188

D

debus 116,156, 179
deliveries 168, 194
demonstrations 40
detailed body search 196
detail leader 33, 107, 116–120, 153
detail personnel 36
detail rotation schedules 33
discreet surveillance 62
dogs 171

drainage pipes 180
dress 25–26, 29, 58
driver 34, 112–116, 118–119, 122
drive throughs 68
drug cartels 37
DUI check points 40
dummy convoy 67
duress code system 193

E

elevators 96–99, 175
embus 115, 152, 156
emergency services 27, 29, 165, 182
escalators 175
ethical considerations 22
evacuation 31, 50, 128, 152, 175, 182
evasive driving tactics 193
expeditor 56, 189
experience 31
explosive ordinance disposal 35, 50, 57, 171, 184
external considerations 39–40, 166

F

false start 67
fax machine 175
female security personnel 169
field of view 161
fire extinguishers 131
fire fighting equipment 27, 170, 186
fire protection systems 182
firing positions 161
fixed point counter surveillance 69–71
flash bangs 119, 121, 131
flat angle of shot 161
flexibility 18, 24, 40, 58, 121
floor plans 180
follow car 36
follow vehicle 114
foot movement 71, 79, 81–90
foot surveillance 64
foreign intelligence agencies 61
frisk 195
funerals 109

G

gang violence 37
garbage 194
gift 177
government 39–40
GPS 177, 197
gunship 36, 113

H

health concerns 22
helicopter landing zones 35, 106
high profile/high visibility 30, 153
high profile vehicles 128–129, 130
high threat environment 16
high threat level 18
holding area 174, 181
holidays 40, 54, 109
hospital 111, 190, 197
hostile nation 37
host nation license plates 193
host nation military 61
hotel 57
hotel management 184
hotel room 27, 186
hotel security 184
hotel survey 184

I

ID badge 57
immigration 26, 188
indirect fire 170–171
information 17
inner protection area 18, 29, 168
intelligence 37
internal considerations 39, 167
interpretation 76
intersections 148–149
itinerary 51, 52

K

Kevlar blanket 127
key personnel data sheets 183
key terrain 178
kill zone 31, 120, 121

L

lane changes 146–147, 148–149
lapel pin 57
law enforcement 29, 182, 194
lead vehicle 36, 113, 114
lighting 180, 183
local law enforcement 27, 40
local staff 169
location reporting 162–163
low profile/low visibility 30, 154–156
low threat level 18
luggage 27, 176, 187

M

mad minute 119
magnetic azimuth 178
main body vehicle 35, 114, 116, 117, 118, 120, 122
maps 176, 177, 197
map study 105
masking activities 77
media 40, 42, 61, 162, 176
medial capabilities 106, 190
medical kits 35, 131, 156
medium threat level 18
metal detection system 195
mid protection area 18
military exercises 40
mindset 23, 25, 90
minefields 108
mission site 178
mobile surveillance 65
motion activated lights 170, 194
motorcade 50, 65, 174
moving roadblock 140

N

night shift 170
nontechnical signal devices 131

O

observation 30, 75, 205–208
observation points 194
observer 161
off duty schedules 34
on-site security 27
operating environment 37, 75
operations center 33
operations orders 199
outer protection area 18, 168
overt profile 153

P

parades 40
passport 56
peeking 65, 207
perception 76
perimeter 166, 180
personal security 16
personal security company 13
personal security detail 11, 29–30, 43
personal security officer 33
physical handicap 22
physical security 16
physical security assessment 44

physiological indicators 77
planning 17, 47, 172
point of contact 52, 174, 182, 190
political climate 39
position security 162
poverty rate 37
practice 31
predictability 47
primary objectives 17
pro-active protection 15
pro-active work 23
proactive security 47
props 205–206
protective detail 35
protective detail 79
protestors 108
protests 40
public speaking 92
public venues 91

Q

quick body search 195
quick reaction force 35, 120, 153, 154, 156–157

R

radio 35
random security checks 169
rapport 58, 185
reactive protection strategy 16
reception line 93–95
reconnaissance 41
rehearsal 31, 42
reinforced bumpers 122
religious pilgrims 109
reporting system 162–164
rescue equipment 156
resources 17, 47
restaurants 50, 187
restraint 31
risk assessment 48
risk management 26, 39, 43–45
risk reduction 15, 44
roof 181
room assignments 184
roundabouts 68, 138
round sheets 168
routes 27, 28, 49, 71, 111, 165
route planning 105–109
route selection 101–103
routine 44, 103
run flats 122
rural environments 64

rural site 165

S

safe haven 106, 111, 113, 119, 120, 174–175, 181, 193, 197
safe room 167
scout 113
scout vehicle 36
searches 157, 193, 195
second-in-charge 34
security advance patrol 35, 69, 113, 153, 155–156
security environment 46
security equipment 182
security operations center 34, 35, 58, 114, 162, 168
security operations room 59, 167, 184
security perimeter 114
security post 114
security room team 34
security vehicle 120
self sealing fuel tank 122
sewers 180
shift leader 33, 34, 50, 115, 121, 152
short notice movement 102
signs of intention 77
site security 49
site survey 114
situational awareness 18
smoke 121, 131
sniper 108, 161
sniper positions 178
sniper rifle 160
SOP 121
spare tires 122, 131
special movement 102
sporting events 40, 109
staging 174
staging area 116, 174, 189
stairwells 175
static site operations 154, 156–157, 165–172
static surveillance 63–64
stationary roadblock 141
steel plate doors 124
subconscious 23
suicide bomer 205
surveillance 41, 61–72, 77, 153, 168, 176, 193
surveillance detection team 29
surveys 173, 197
sustained areas of fire 109
sweep 114

T

tactical leader 33
tactical mindset 78

target 41
target hardening 47
target selection 41
tarmac 189
technical security measures. 182
technical surveillance 66
telephones 167, 175
temporary site security 172
terrorism 37
threat 37–42, 46, 61, 117, 155
threat assessment 44, 46
threat level 18, 55, 122, 187
threat operations cycle 41–42
tinted windows 155
traffic 197
traffic circle 68, 138
traffic light 68
training schedules 34
trash 167
TSCM sweeps 184

U

unscheduled stops 67
urban site 165

V

vaccinations 55
vehicle armor 120
vehicle checklist 113
vehicle control points 108
vehicle movements 30
vehicle obstacle 121
vehicle reception 139, 150–152
vehicle recovery equipment 131
vehicle registration 193
vehicle service 106
vehicle tracking system 66, 193
VIP lounges 26, 188
visa 56
visitors 169
vulnerability assessment 44, 47

W

walking routes 174
warning order 199
war zones 107
weapons 29, 54, 56, 90
weather 40, 54, 106

FOR MORE INFORMATION ON LAW ENFORCEMENT AND SELF-PROTECTION TRAINING BOOKS AND DVDS:

WWW.TURTLEPRESS.COM